ELECTRONIC
GENIE

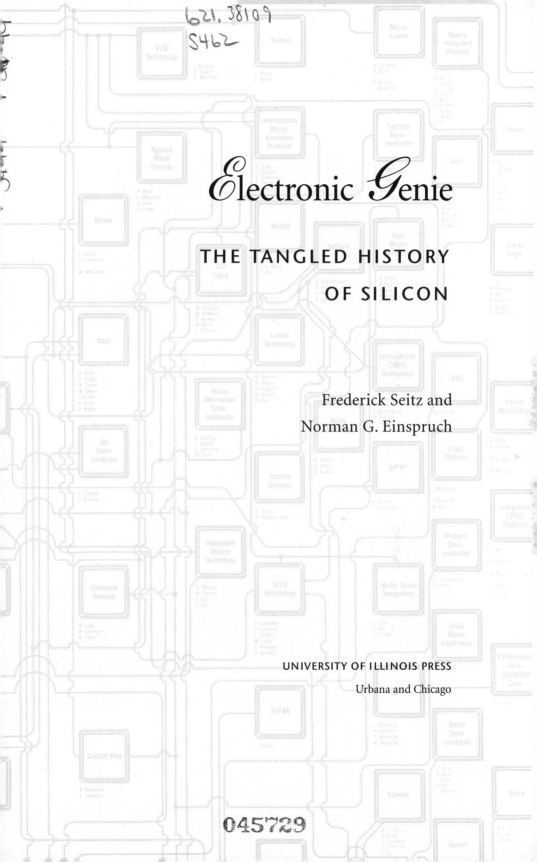

Electronic Genie

THE TANGLED HISTORY

OF SILICON

Frederick Seitz and

Norman G. Einspruch

UNIVERSITY OF ILLINOIS PRESS

Urbana and Chicago

Frontispiece: Antoine Laurent Lavoisier and Madame Lavoisier.
Painted by Jacques Louis David, 1788. (Courtesy of the
Metropolitan Museum of Art, New York City, and Mr.
and Mrs. Charles Wrightsman.)

Publication of this book was supported in part by grants from the
Alfred P. Sloan Foundation, the Ogden Corporation, and the Texas
Instruments Foundation.

Designed and typeset by B. Williams & Associates. The text is Adobe
Minion, designed by Robert Slimbach; the display is Stone Sans,
designed by Sumner Stone, with Shelley Andante swash caps.

This book is printed on acid-free paper.

LIBRARY OF CONGRESS CATALOGING-IN-PUBLICATION DATA
Seitz, Frederick, 1911–
    Electronic Genie : the tangled history of silicon /
Frederick Seitz and Norman G. Einspruch.
        p.   cm.
Includes bibliographical references and index.
ISBN 0-252-02383-8 (cloth)
1. Solid state electronics—History.   I. Einspruch, Norman G.
II. Title.
TK7809.S45   1998
621.381'09—DC21   97-21145
                                              CIP

CONTENTS

To the memory of

JOHN BARDEEN (1908–91)

WORKING WITH Walter H. Brattain at the Bell Telephone Laboratories, John Bardeen made evident, through experimental demonstration and
during a relatively brief period of time, the nature of the obstacles that had
blocked previous attempts to develop the field-effect transistor and thereby
opened the door to its practical realization. Simultaneously, he and Brattain
invented the bipolar minority-carrier injection transistor, the first practical
operating transistor and the progenitor of the semiconductor devices and
circuits that followed. This invention is having as profound an effect upon
the course of civilization as any that preceded it.

Working with Leon N. Cooper and J. Robert Schrieffer at the University
of Illinois, where he spent the great majority of his professional career,
Bardeen led the theoretical program that disclosed the origin of low temperature superconductivity, an open mystery since its discovery by Kamerlingh Onnes in 1911.

For these monumental advances, John Bardeen and his colleagues were
awarded Nobel prizes in 1956 and 1972.

PREFACE

$\mathcal{T}$HE ORIGINS OF THIS BOOK are almost as complex as the subject it covers, namely, the way in which elemental silicon came to play a major role in the field of electronics.

In 1993, Charles C. Torrey, who had been coordinator of the research and development program on semiconducting diodes at the Radiation Laboratory of the Massachusetts Institute of Technology during World War II, was asked to give a lecture on the program he had led there. The presentation was to be given at a meeting of the American Physical Society in Pittsburgh in March 1994. Torrey was somewhat ill at the time and proposed that Frederick Seitz, who had been involved in the program essentially since its inception, be an alternate speaker. Seitz agreed with the provision that Torrey review, edit, and supplement the text prepared for the lecture.

The text of the presentation was published in *Physics Today* in January 1995. The printed version prompted several responses, including a friendly critical one from Robert V. Pound, who had also been at the Radiation Laboratory. Pound pointed out that the work of Denis M. Robinson and his colleagues, who had introduced the use of silicon diodes for heterodyne mixers in microwave radar in England early in World War II, was not even mentioned in the article. Although Seitz was well aware that the English had started using silicon diodes, he knew nothing of Robinson or his contribution. With leads provided by Pound, he decided to fill the gap and submit a letter to *Physics Today* that described Robinson's contribution.

The available literature soon revealed the fact, however, that Robinson's work had been inspired by something he had read in the German literature. That material was, in turn, inspired not only by work in the field of microwave technology in the 1930s, as well as that involving the use of crystal diodes in the early days of radio in the 1920s, but also by very early research in the field of coded wireless at the beginning of the century. As Seitz delved into the historical background, he was soon aided by an essentially worldwide network of friendly advisers and contributors. Information was provided by individuals in countries ranging from Japan to Russia, with major revelations from England, France, Germany, and Italy as well as the United States. A brief summary, somewhat in the nature of a resource paper, since it included extensive reference material, was published in the *Proceedings of the American Philosophical Society* in 1996. It focused on developments that extended from the first use of semiconducting diodes in coded wireless, starting in 1901, through the use of silicon diodes as heterodyne mixers in radar in World War II.

Left open, however, were several broad and important questions: How did crystalline forms of the element silicon happen to be available to wireless experimenters at the time when wireless telegraphy was still at the experimental stages? How did it happen that the Bell Telephone Laboratories focused on the development of a semiconducting triode immediately at the end of World War II? Was it a matter of chance or the result of a premeditated program? What was the precise sequence of events that led to the invention of the first practical transistor? How did it happen that the invention and development of the integrated circuit was carried through by individuals involved in what might be termed startup organizations, at least as far as basic work in electronics is concerned, rather than by people in companies that had previously been leaders in vacuum tube electronics and had much to gain or lose as a result of a revolutionary new development?

The present book has been written with the intention of filling out as much of the broader picture as is possible in a relatively small volume. In brief, it covers the developments from the time of Lavoisier, who first predicted the existence of silicon as an element, to the present time, when that element has become a revolutionary agent in the conduct of everyday affairs as well as in the technological world.

Part of the pleasure of carrying out the research underlying this book has involved bringing key figures from the past back to life—a pleasure fully matched by the complementary one of reviewing recent developments with a number of individuals who have worked at the frontier of semiconductor electronics in recent decades. In truth, one might say that this book has had at least a hundred authors. To the extent possible, they and their contributions are acknowledged in the following pages.

It is also a privilege and pleasure to state that the book would not have been possible to produce without the generous support of the administration of the Rockefeller University and Mrs. Florence Arwade.

ACKNOWLEDGMENTS

$\mathcal{I}$T IS A PLEASURE to express gratitude to the many individuals who have contributed to this volume. Credit is also given to them, as well as to others, in the text of the book.

This historical record developed in three successive stages. The first, as described in the preface, centered on the exploitation and improvement of silicon-tungsten diodes for use as heterodyne mixers during World War II in connection with the work of the Radiation Laboratory at the Massachusetts Institute of Technology. A succinct account of this aspect of the use of silicon in electronics was published in *Physics Today* in January 1995. Much of that account was derived from material found in *Crystal Rectifiers,* by Henry C. Torrey and Charles A. Whitmer, volume 15 of the Radiation Laboratory Series, as well as from the recollections of Torrey and Frederick Seitz. Torrey had administered the semiconductor program at the laboratory. In retrospect, the article owed so much to Torrey that his name undoubtedly should have been added as one of the authors.

The second stage covered developments beginning with the innovative experiments of Heinrich Hertz in the 1880s, which led to wireless telegraphy, and extending to the period of microwave and radar research prior to the start of World War II. The research dealing with that period depended upon contributions from a much larger group of individuals. The study was initiated when Robert V. Pound of Harvard University, who had served on the staff of the Radiation Laboratory, called attention to the seminal work on semiconductor diodes in England by Denis M. Robinson very early in the war. As is emphasized in the paper in the *Proceedings of the American Philosophical Society,* mentioned in the preface, the historical research involved major contributions from Pound as well as from John M. Anderson of Scotia, New York, and formerly of the General Electric Research Laboratory; Berthold Bosch of the Ruhr University at Bochum; Louis Brown of the Carnegie Institution in Washington, D.C.; the late John H. Bryant of the

University of Michigan; Robert W. Cahn of Cambridge University; Lillian H. Hoddeson of the University of Illinois; Kenneth G. McKay, formerly of the Bell Telephone Laboratories; C. Marcus Olson, formerly of the DuPont Company; Helen Samuels of the Archive Center of the Massachusetts Institute of Technology; and Harald D. Robinson of Cambridge, Massachusetts, Denis Robinson's son. These individuals essentially followed the work from start to finish. Berthold Bosch was particularly generous in providing duplicates of many items of historical and technical interest from his extensive personal files. Robert Cahn was unstinting in searching out special items from British sources.

Additional significant contributions were made by Pierre Aigrain of Paris, France; Giuseppe Baldacchini of the ENEA Center at Frascati; Franco Bassani of the Scuola Normale Superiore of Pisa; Robert Buderi of Cambridge, Massachusetts, who has since published a book on the history of radar; Gianfranco Chiarotti of the University of Rome; Erich Hahne and G. Barthau of the University of Stuttgart; Jack S. Kilby, the inventor of the integrated circuit; James S. Koehler of the University of Illinois; Michael Riordan, the science historian; and Heinz von Foerster, formerly of the University of Illinois.

The third phase of the study, which led to this book, was initiated out of a desire to achieve three goals: First, to understand more fully the circumstances that made crystals of elemental silicon readily available to experimenters in the early days of wireless telegraphy; second, to cover the transition from the development and use of crystal diodes in radar to the invention of the discrete transistor, the integrated circuit, and the single chip microcontroller (or microprocessor)—along with a view toward the future; third, to fill in additional details of the developments that took place during the first two stages of the study.

Many of the individuals who were of great help in connection with the preparation of phase two of the study, and whose names are listed in preceding paragraphs, continued to follow the work and contribute to it. In this connection, particular mention should be made of the help from Pierre Aigrain, Berthold Bosch, and Robert Cahn. In addition, we owe special gratitude to Zhores Alferov, vice president of the Russian Academy of Sciences and director of the Ioffe Physical-Technical Institute in St. Petersburg; Robert M. Ehrenreich of the National Research Council of the National Academy of Sciences; S. T. Harris, J. S. Kilby, James Comfort, Esq., and Harold Levine, Esq., of Dallas, formerly of Texas Instruments, Incorporated;

Turner E. Hasty of Dallas and formerly associated with Texas Instruments and SEMATECH; Lillian H. Hoddeson and Nick Holonyak of the University of Illinois; Howard R. Huff of SEMATECH; R. A. Laudise of the Bell Telephone Laboratories of AT&T; Norman P. Neureiter of Texas Instruments; Leslie Reynolds, librarian at the University of Illinois; Joel Shurkin, who is preparing an extended biography of William B. Shockley; Charles Susskind of the University of California at Berkeley; and Charles A. Wert of the University of Illinois. Turner Hasty and Howard Huff read through early versions of the complete text and made very valuable scientific, technical, and editorial comments that influenced the book.

Much effort was devoted to chapter 14 in an attempt to understand the sequence of events leading to the invention of the several forms of discrete transistor. We are particularly indebted to Dr. Ian M. Ross, former president, and William F. Brinkman, current vice president for physical research, of the Bell Telephone Laboratories, for valuable discussions of the various stages of development. We are similarly indebted to Nick Holonyak as well as to Lillian H. Hoddeson and Michael Riordan. Through the courtesy of the latter two, we had the privilege of obtaining an advance copy of their book *Crystal Fire* (New York: Norton, 1997), which provided valuable source material concerning the invention of the discrete transistor. We appreciate the cooperation of E. Barber of the Norton Publishing Company in this regard.

We are grateful to Berthold Bosch for introducing us to Herbert Mataré, with whom we have had important exchanges of information, including that concerning the discovery of transistor action in polycrystalline germanium, which is described on page 174 of this book.

We benefited much from interviews with the late Jerry R. Junkins, former chief executive officer of Texas Instruments, and with Gordon E. Moore, the chairman of the board emeritus of Intel Corporation. In this connection, we would like to express gratitude to Lillian H. Hoddeson for the privilege of reading the text of an interview she and Michael Riordan had with Gordon Moore. It provided valuable insights into the period in which Moore was a major participant, as a chemist, in Shockley's company in Palo Alto.

At the time the book was being planned, we had useful discussions concerning the outline and general structure of the text with Jeffrey L. Sturchio, who was then editor of the History of Modern Chemical Science series, issued under the auspices of the American Chemical Society and the Chemical Heritage Foundation, in which a book by one of us had been published previously. His successor as editor, Anthony S. Travis, subsequently made many important editorial suggestions for improving an early version of the text, for which we are most grateful.

In the course of writing the book, we were asked by Dr. Probir K. B. Bondyopadhyay, an editor of the *Proceedings of the Institute of Electrical and Electronics Engineers,* to prepare an article for a special issue of the *Proceedings* commemorating the invention of the transistor. We are indebted to him for valuable exchanges of information and viewpoints that have influenced portions of this book, particularly chapters 2 and 14.

The problem of finding suitable illustrations to go with a text is seemingly never-ending. We are particularly indebted to the Metropolitan Museum of New York City for permission to use the David portrait of the Lavoisiers as a frontispiece.

Many illustrations came from relatively traditional sources such as the archives of the American Institute of Physics; the Center for the History of Electrical Engineering of the Institute of Electrical and Electronics Engineers, Rutgers–The State University of New Jersey; the National Academy of Sciences; and the Smithsonian Institution. We express thanks to Tracy Keifer, Andrew Goldstein, Janice Goldblum, and Mark Rothenberg, respectively of those institutions, for their help. Ms. Jane Colihan, picture editor of the *American Heritage of Invention and Technology* magazine of Forbes, Incorporated, was especially helpful, in cooperation with Ms. Keifer, in locating a photograph of Lars Grondahl. The Deutsches Museum in Munich and the Academie des Sciences in Paris were also valuable general sources. The office of the Royal Society of London was cooperative in steering us to the sources of the photographs that appear in the memoirs of its members. We express appreciation to the staff there for its help. It should be added that Robert Cahn provided special aid in some of these cases, particularly in obtaining images from the Lotte-Meitner Graf studios and in the subsequent search for a photograph of Johann Königsberger. The Stanford University Archives generously offered a historical photograph of one of the first klystrons, along with its inventor and his colleagues. The archivists at the Edison National Historic Site of the National Park Service not only gave us a photograph of Edison taken in Washington, D.C., when he was demonstrating his newly invented phonograph, but also described the circumstances under which the picture was probably made.

The photograph of Ferdinand Braun was obtained from the University of Tübingen with the special help of Professor H. Metzner and Fritz Seitz. Zhores Alferov furnished us photographs of Dmitri Mendeleyev, Alexander Popov, and the Soviet radar group that produced a remarkably effective multicavity magnetron in the late 1930s. The Science and Technology Museum in London allowed us to use an early photograph of J. C. Maxwell, which we had not seen previously. Jack S. Kilby, in addition to giving us one of his photographs, loaned us an important document issued by the U.S. Army Signal Corps in 1918. We have included a photograph of its cover in the book. Georg Busch not only gave us a copy of his photograph, but he and his colleague Pierre Cohn helped us find some of the photographs that appeared in his comprehensive review of the history of semiconductors. The University of Jena then generously gave us a copy of the photograph of Karl Baedeker. Berthold Bosch, in turn, sent us a copy of a photograph of Hans E. Hollmann, taken from a discontinued journal. Pierre Aigrain furnished a photograph of Maurice Ponte, along with an account of some high points of French magnetron research.

Several photographs, including that of Denis Robinson, were taken from the illustrated book *Five Years at the Radiation Laboratory*, originally issued by the Massachusetts Institute of Technology, Cambridge, Massachusetts, in 1946 and reprinted with some expansion by the Institute of Electrical and Electronics Engineers (IEEE) for the 1991 International Microwave Symposium. We are grateful to John Bryant and his colleagues for a copy of this. In parallel, some of the photographs were taken from Torrey and Whitmer's volume devoted to semiconductor rectifiers, *Crystal Rectifiers*. Herbert Friedman was very generous in helping us gain access to the archives of the Naval Research Laboratory, just as William O. Baker and R. A. Laudise helped us to gain access to the archives of the AT&T Bell Laboratories. We are indebted to Henry Apfelbaum and his colleagues there. Lillian H. Hoddeson shared with us a relatively rare photograph of R. S. Ohl. Arnold Tubis of the physics department of Purdue University gave us photographs relevant to the wartime research on germanium carried on there. Mrs. Katherine M. Hurd went through family collections and loaned us material dealing with the period in which her grandfather Theodore Vail was extending the geographical range of the AT&T system nationwide with the help of an improved De Forest triode. Nick Holonyak shared valuable material from John Bardeen's papers, including diagrams that Bardeen had used in his lectures. Turner Hasty, among many other things, opened the door to some of the photographs available in the archives of SEMATECH. Madame Marion Tournon-Branly not only gave us one of her favorite photographs of her grandfather, but served as guide for a tour of the Branly Museum in Paris.

Gary Boone, who has received recognition from the U.S. Patent Office as the inventor of the microcontroller, gave us, on request, a copy of his photograph. Similarly, David Pines and the University of Illinois presented us with a photograph of John Bardeen and his colleagues taken at the Nobel ceremony in 1972.

We are grateful to a number of private companies, in addition to AT&T, for donating important photographs. Among these are: Forbes, Intel, Materials International, Microsoft, Next Software, Robertson Stephens, Synaptics, and VSLI. Sharon Bittle of Intel was additionally helpful in many ways. We are grateful also to the archivists at Texas Instruments for providing photographs and permitting us to copy diagrams from the series of technical books the company published in conjunction with the McGraw-Hill Book Company. Not least, we appreciate the assistance of H. Fusstetter of Wacker Siltronic in supplying the fine photographs of silicon that appear on page 249.

Finally, we wish to thank the staff of the University of Illinois Press for its splendid cooperation in the preparation of this book, along with much productive advice.

ELECTRONIC
GENIE

ONE

ROOTS

$\mathscr{T}$HE SO-CALLED INFORMATION or computer superhighway is paved with chips of crystalline silicon. This is a triumph of advances in the understanding of solid-state or materials science. It is also a product of the knowledge gained in the convergence of major areas of chemistry, metallurgy, and physics, particularly those related to the behavior of solids in the presence of electric and magnetic fields, when applied to the design of electric circuits.

The challenges presented by evermore demanding applications of sophisticated chemistry and the growing need for more intricate circuits have continued to be the most pressing, once the basic physical principles are understood and set in place. A modern factory for producing silicon chips is a masterpiece of automated chemical engineering, operating most efficiently and effectively when the production area is populated by the fewest number of operators possible, in order to minimize contamination and to improve both control and uniformity of production processes. Moreover the evolution of increasingly more powerful computers has hinged on advances in the design of circuits that match advances in sophistication of chip production.

It is thus only reasonable to ask how this has happened, that is, to construct the history of the developments that led to the present-day predominance of silicon in electronics.

This book deals in substantial part with a segment of the history of technology, and not with science of the fundamental or exploratory type. The latter tends to be well documented in scientific journals for purposes of communicating scientific information as well as establishing personal priorities for discoveries. In contrast, the individuals involved in practical technical advances, that is, in applied science and engineering, usually lead far more complicated, and certainly no less stimulating, lives than academic scientists. They deal with assignments for which the outcomes are usually prescribed and with patents that protect their organizations' intellectual property rights and may be tied to some level of industrial or government secrecy. Good manufacturing processes may be retained as trade secrets and not disclosed to the public in any form. Patent protection involves licensing, lawyers, entrepreneurs, bankers, and legal regulations.[1] Moreover, as we shall observe frequently, closely parallel inventions may be developed and patented in different countries in somewhat different forms, which adds to the complexities of gaining knowledge concerning origins and priority.

CHEMISTRY

The beginning of the story is probably best tied to the discovery of elemental silicon early in the nineteenth century by a group of chemists following in the footsteps of the founder of modern chemistry, who had completely revolutionized the field during the previous half century. He was Antoine Laurent Lavoisier,[2] an individual with a remarkably broad range of professional talents, including knowledge of economics and management of public enterprises in the royal government. (See frontispiece.) With almost a single stroke, he demolished concepts of the constitution of matter and the basic elements of which it is composed that had evolved over two and a half millennia and introduced the modern concept of basic, supposedly immutable constituents. Presumably they were finite in number, but much larger than the ancient notion of earth, water, air, and fire.

The path that Lavoisier followed was a stumbling one since there was much to disentangle. He did demonstrate, however, that many ordinary substances familiar to the chemists and alchemists of the day were composites formed by the combination of more basic elements, particularly metals, with oxygen and sulfur.

At a critical point in his analysis, Lavoisier was greatly influenced by the study of the physical and chemical properties of gasses carried out by the clergyman-chemist Joseph Priestley.[3] The French scientist was particularly interested in the reactive properties of pure oxygen, which Priestley had obtained in 1774 from the thermal decomposition of mercurous oxide, and in Cavendish's earlier decision that hydrogen was a primary material, that is, an element in modern terms. There is little doubt, however, that Lavoisier's vision of chemistry was deeper and broader than that of any of his contemporaries. He appreciated the fact, whether or not he thought in such terms, that the basic chemical elements represented something in the nature of a finite number of playing cards in a deck that could generate many "hands," or compounds, in a given situation, and that the deck could be employed in many different types of application.

It is worth adding that Lavoisier's widow married the American-born English physicist Benjamin Thompson, after Lavoisier was guillotined during the French Reign of Terror. (It was Thompson who first proposed the theory of the mechanical equivalent of heat.) Moreover, Priestley's life also was marked by the events in France. In the same year that Lavoisier was executed, Priestly had to flee England for America because of his open sympathy with the French revolution (he actually became a French citizen soon after the revolution started).

Having generated or reduced several oxides, Lavoisier had suggested in 1789 that quartz, or silica—that is, silicon dioxide—was probably the result of the union of oxygen with an as yet unidentified but highly important element. Since his own life was cut short, it was left to others to pursue this challenge.

Apparently the first marginally successful preparation of elemental silicon was achieved by Joseph Louis Gay-Lussac and Louis J. Thénard, who passed silicon fluoride over heated potassium in 1811. However, a much more conclusive experiment was carried out by the Swedish chemist Joens Jacob Berzelius in 1824 using an alternate method based on the reduction of potassium fluosilicate with metallic potassium. The element was then recognized to be one of the most abundant in the lithosphere, that is, the crust of the earth. Even at that early stage, a debate developed as to whether silicon was a metal or an insulator, Berzelius taking the view that it is a metal and Humphrey Davy that it is an insulator. This issue was not really resolved with certainty until World War II, along with deeper theoretical understanding of crystalline solids.

The scientific study of silicon continued to advance during the nineteenth century. For example, Friederich Wöhler developed definitely crystalline forms in 1856 and 1858. Although the chemistry of compounds involving silicon as a major constituent, not least the silicates, occupied a central position in the field of inorganic chemistry through much of the nineteenth century, little seems to have been said in the literature about the elemental crystalline form until it became of interest to the metallurgists during the last quarter of the century.

THE PERIODIC TABLE: MENDELEYEV

The climax to the revolution in chemistry that Lavoisier had started occurred in 1871 with the creation of the periodic table of the elements by the Russian chemist Dmitri I. Mendeleyev. Lothar Meyer had previously observed that certain triads of elements, such as lithium, sodium, and potassium, and magnesium, calcium, and strontium, exhibit similar chemical properties. Following leads of this type, Mendeleyev gave each known element a position in a single grand ensemble and identified gaps that he predicted would be filled with as yet undiscovered elements. These undiscovered elements included gallium, scandium, and germanium. Germanium, which will play a significant role in this story, is in the same family as carbon and silicon. In fact, under different circumstances the emphasis in this book might have been given to germanium instead of silicon. However, germanium possesses several disadvantages relative to silicon in electronic applications: It is much less abundant, being present in the lithosphere only to the same extent as some of the lesser rare earth elements; its electrical and chemical properties are much more sensitive to temperature and other influences than those of silicon; and, most important, its oxide, being soluble in water, does not possess the same passivating and insulating properties as silicon oxide.

Mendeleyev's table served the needs of the chemistry community exceedingly well. Moreover, it took on additional meaning with the development of the quantum version of atomic mechanics.

METALLURGY

The rapid pace of industrial development in the second half of the nineteenth century led to a great expansion in the production of iron for both commercial and military purposes, with much emphasis placed upon steel of controlled carbon content for such applications as machinery and tools, rails for the burgeoning railroad system, beams for bridges, and, of course, for weaponry. Metallurgy had been a highly developed, empirically grounded field of technology that was securely in the hands of experts who often applied their art with what were regarded as proprietary skills. However, it could not escape the influence of expanding scientific developments, least of all the instrumentation of physics and the ideas of chemistry. Apart from the introduction of the microscope and the analytical spectroscope, the greatest influence emerged as a result of testing the effects of hitherto untried alloying agents upon the properties of metals, particularly carbon steels. By the 1890s the age of alloy steels was in full swing. Many physical chemists rapidly found a productive home in the metallurgical laboratory as they applied inorganic chemistry, quantitative analysis, thermodynamics, phase diagrams, and experience with the microscope.

It was observed in the process that the addition of elements such as nickel, chromium, manganese, and even silicon to iron in controlled amounts could impart improved strength, toughness, ductility, or even resistance to corrosion.[4] In this way, the elemental form of silicon began to lose its "academic" status in the 1890s and become an important part of the metallurgist's stock-in-trade. In one sense it could be looked upon as a poor man's alloying agent because the element was inherently abundant.

Soon there was a strong interest in producing silicon in the pure forms suitable for various metallurgical purposes. For example, H. N. Warren found that it could be produced economically and in quantity by heating a mixture of silica, carbon, and iron. Similarly, E. Vigouroux obtained a useful product by the reduction of silica in the presence of magnesium and zinc. Many other imaginative experiments were carried out in the industrial world at the time with the goal of producing reasonably pure silicon. Eventually, the Carborundum Corporation developed a process that depended upon the reduction of the purest available silica in the form of sand with coke—a product that satisfied the immediate needs of the steel industry. Perhaps to avoid possible infringement of patents based on other processes, the German industry developed a process based on the use of silicon carbide and coke as starting materials.

One particularly interesting byproduct of the work with silicon-iron alloys was the discovery that there are circumstances in which they possess advantages in the preparation of magnetic sheet for transformers and other devices. In some cases, for example, the rolled sheet would exhibit preferred orientation of the constituent crystals, or "grains," and in such a direction that the energy losses associated with repeated magnetic reversal (so-called hysteresis losses) were lower than they would be for unalloyed specimens. Prime movers in this type of research at the turn of the century were Robert A. Hadfield in England and E. Gumlich in Germany.[5]

Once silicon became readily available at reasonable cost, it found many other uses. For example, it was commonly added in appropriate quantities to ladles of molten steel to combine with dissolved oxygen. This prevented formation of voids in ingots through the release of bubbles of carbon monoxide or dioxide during solidification. Moreover, silicon, along with other elements such as magnesium, eventually became incorporated into aluminum alloys used for special purposes, such as piping or parts of mechanical structures where enhanced strength is desirable.

CRYSTAL SYMMETRY

Natural crystals, sometimes of gem quality and with their regular structure, have attracted the attention of the curious since the dawn of history. Of very special interest was the fact that they could exhibit different physical properties, such as a refractive index for light, thermal conductivity, or elasticity, in different internal directions; that is, they can exhibit anisotropy. As a result, it was only natural, once the scientific movement had gained momentum in Europe in the eighteenth century, that individuals with a geometrical bent would begin to examine the geometrical regularity and intrinsic symmetry of single crystals to determine whether the symmetry is that of a cube, a hexagonal or rectangular prism, or perhaps of much lower order. It was eventually found that the types of symmetry actually occurring in nature are limited in number to thirty-two so-called classes or groups.

One of the first systematic investigators was René Just-Haüy, who, in the latter part of the eighteenth century, studied the regularity of the angles of cleavage of different specimens of the same mineral, as well as other aspects of symmetry. He found that such cleavage angles were fixed characteristics of the mineral. A successor, August Bravais, concluded in a study made in 1850 that crystals must be composed of regularly stacked, three-dimensional arrays of small, identical, geometric units, or "cells," having the same symmetry as the actual specimen. In other words, he concluded that the specimens are composed of a three-dimensional lattice of such cells in the ideal crystalline state. In fact, he demonstrated that such a regular lattice structure would readily explain the limitations placed on the symmetry types or groups actually found in nature.

By 1890 three individuals, W. Barlow (England), E. S. Federov (Russia), and A. Schoenflies (Germany), had independently worked out the symmetry properties that might be expected to occur in three-dimensional lattices if one includes translational symmetry. They thereby laid the groundwork for present-day crystallography based on the use of X-ray, electron, or neutron diffraction. They found that there were 230 possible symmetry patterns, or space groups.

PHYSICAL PROPERTIES OF CRYSTALS

Apart from such studies of geometrical relationships, there was also much interest in measuring the mechanical, thermal, optical, and electrical properties of such crystals. Indeed, this became a central topic of physics and chemistry through much of the nineteenth century and has been investigated with increasing sophistication ever since. It is notable that Antoine H. Becquerel was engaged in research of this nature when he discovered the natural radioactivity of uranium in 1896. Moreover, until the discovery of radioactivity, the field of crystal physics and chemistry was of major interest to the brothers Jacques and Pierre Curie, the latter of whom later discovered radium with his wife, Marie. One of the major contributions of the brothers to crystal science was the discovery in 1880 of the piezoelectric effect, whereby certain crystals having less than the highest permissible levels of symmetry develop a voltage across or along specimens having a specially selected orientation upon compression or shear. Conversely, such crystals change dimensions on application of an electric field. Quartz crystals possess this property and have come to play an important role in the electronics industry because of both this property and their inherent chemical stability. They can be used as electromechanical (piezoelectric) resonators and serve as stabilizers of oscillating electric circuits, such as those found in computers and quartz watches. Today, most such quartz resonators are fabricated from large, highly refined synthetic crystals.

An interesting, almost encyclopedic, summary of macroscopic crystal physics was first published by Woldemar Voigt in 1914.[6] It was reissued in 1928 with additions by Max von Laue, who, in 1913, had discovered that crystals diffract X rays. This revelation in turn opened the doorway to X-ray spectroscopy and to the determination of the arrangement of atoms in both crystals and constituent molecules—procedures that have been augmented with the use of both electrons and neutrons and have more recently proven to be invaluable in the determination of the structures of large biochemical molecules.

CRYSTAL RECTIFICATION

Another brilliant and diligent worker in portions of the field of crystal physics was Ferdinand Braun, who plays a prominent role in this story (see chapter 2).[7] In 1874, Braun, at the start of his career, studied the electrical conductivity of a number of poorly conducting natural sulfide crystals. He was surprised to find that they did not appear to have the same electrical conductivity for both forward and reverse directions of the current. The "rectification" effect seemed to represent a violation of Ohm's law describing the linear relationship between current and applied voltage found in good electrical conductors. Braun noted, however, that the effect depended on the relative size of the electrodes, being greatest when one electrode was considerably smaller than the other. Doubt was cast upon his results by others who had difficulty reproducing the effect after it was announced, but he succeeded in replicating and confirming his results. They remained a part of the background scientific literature until the end of the nineteenth century, when they had a major influence on a new branch of technology, namely, wireless telegraphy. Subsequent research showed that the current-voltage relationship inside homogeneous specimens of such materials (or semiconductors, as they have come to be called) does indeed conform to Ohm's law. The rectification is associated with what has come to be called a surface "blocking layer" (see chapter 5).

MAXWELL'S EQUATIONS

In 1865, James Clerk Maxwell developed the first unified form of the equations of electromagnetism, one of the great achievements of the century. The equations were based on the following principles: The inverse square law of attraction and repulsion of unlike and like electrostatic charges (Coulomb's law); the continuity of magnetic fields and the absence of free magnetic poles; Michael Faraday's and Joseph Henry's law of induction of electric fields by a changing magnetic field; and Ampere's law governing the magnetic field that encircles a current-carrying wire. In order to include the presumed generation of a magnetic field by varying the electric field of a moving charge, Maxwell extended Ampere's law by adding a hitherto absent term concerning the relationship. Finally he examined both general and special solutions to the newly formulated ensemble of four equations.

Among the solutions, he discovered that the equations permitted the transmission through space, including a vacuum, of what are now termed electromagnetic waves. Moreover, when appropriate electric and magnetic parameters were included, the velocity of propagation of the waves coincided with the known speed of light. He concluded that ordinary visible light consists of such waves of very short wavelength, of the order of a micron, that can be generated by matter under appropriate conditions of excitation.

Unfortunately, Maxwell died of cancer in 1879 at the age of forty-eight and did not live to see the emergence of the age of wireless in the hands of his successors.

HERMANN VON HELMHOLTZ AND HEINRICH HERTZ

Hermann von Helmholtz was one of the broadest and most perceptive scientists of the nineteenth century. Although his greatest interests as a youth had been in chemistry and physics, he had grown up in relatively impoverished circumstances and had to accept an essentially free education in medicine in order to enter professional life. This start did not, however, prevent him from staying abreast of developments in other fields of science. Moreover, his own experimentation was truly interdisciplinary, combining knowledge in physiology, chemistry, and physics. He made great advances in the understanding of the operation of the human ocular and auditory systems.

On learning of Maxwell's new formulation of the equations of electromagnetism and the implications for the theory of light, he wondered if it would be possible to generate electromagnetic waves of arbitrary wavelength using what might be called standard electric components. He offered a prize to anyone who might generate such waves in the laboratory—the prize being offered for a finite period of time (it expired). In parallel, he urged his brightest and most competent advanced student-colleague, Heinrich Hertz, to carry out an investigation of the possibilities.[8] Hertz took the challenge seriously, but at a pace that was determined by his other commitments and his own view of the problem. To begin with, he recast Maxwell's work into a form that fitted in with his own theoretical style and understanding and then undertook a brilliant series of experiments, developing equipment that combined elegance and simplicity. By 1887 he was routinely producing electromagnetic waves with wavelengths in the one meter range. His waves were generated by oscillations of current flow in a capacitively charged resonant dipole circuit containing a radiating antenna and triggered by an arc. The dipole system was charged rapidly by the voltage produced in the secondary windings of a transformer by a collapsing magnetic field. About 15 percent of the energy stored in the dipole system was emitted in the form of radiation, the remainder being dissipated in the circuitry. Auxiliary measurements involving standing waves along a wire demonstrated that the speed of propagation was, as expected, that of light.

Hertz also created waves in the centimeter range of wavelength and demonstrated that they were refracted by prisms made of pitch, much as light is refracted by a glass prism.

To cap the matter, he observed that the flash of light from his arc could trigger the breakdown of other highly charged systems. He pursued this discovery and found that the effect could be prevented by placing an optical shield around the system that was accidentally triggered. Moreover, he demonstrated that the effect was caused by the ultraviolet component of the initiating arc and concluded that radiation of short wavelength was inducing the emission of charge from the metal components of the system exhibiting accidental breakdown. He had accidentally discovered the photoelectric effect. This important discovery was soon followed up by W. Hallwachs, apparently with Hertz's support. The great skill with which Hertz combined theoretical and experimental talents and the thoroughness with which he carried out his observations parallel those later displayed by Enrico Fermi. Unfortunately he died, apparently as the result of an infected tooth, in 1894 at the early age of thirty-seven. Hertz would undoubtedly have been one of the first physicists to win a Nobel Prize.

Although he devised a method for producing electromagnetic waves, Hertz probably would not have taken the lead in developing their use for wireless telegraphy had he lived, granting that he could have become a valued consultant to those who did. If fate had been kinder, he would undoubtedly have been greatly excited by Planck's discovery of the quantum of action imbedded in the physical constant that bears Planck's name, and perhaps even more by Einstein's 1905 papers dealing with the implications of the photoelectric effect and with the special theory of relativity.

HENRY ROWLAND

The brilliant American physicist Henry A. Rowland, probably best known for his creation of machine-ruled optical diffraction gratings but who also made great technical contributions to telegraphy, spent 1875 with Helmholtz in Berlin just before joining the faculty of the Johns Hopkins University, then newly formed.[9] At Helmholtz's suggestion, he took on and successfully demonstrated in a remarkably short time that electrically charged bodies moving at a high speed actually carry a magnetic field with them, in keeping with Maxwell's extension of electromagnetic relationships. He repeated the experiment on two occasions during the next quarter century, in response to critics of the results obtained in Helmholtz's laboratory. In the process, he removed any doubts about his measurements.

Joseph Priestley, who produced pure oxygen in 1774 by the decomposition of mercurous oxide. The availability of oxygen produced in this way permitted Cavendish, who had decided previously that hydrogen is a "distinct" substance, to demonstrate that water is an oxide of hydrogen. (Courtesy of Robert Hale LTD, London.)

Benjamin Thompson, also known as Count Rumford as a result of his special technical services in Bavaria. He was born in America but left for England at the time of the revolution. Working in the Munich arsenal he noted the correlation between the heat produced and the amount of work done while drilling the bores of cannon and concluded that heat and energy were related. (Courtesy of the Deutsches Museum, Munich.)

Joens Jacob Berzelius, the great Swedish chemist who produced elemental silicon in 1824. It remained a chemical curiosity for about sixty years, until the metallurgical chemists took an interest in alloy steels. (Courtesy of Deutsches Museum, Munich.)

Friederich Wöhler, who succeeded in producing crystalline silicon in the 1850s. (Courtesy of the Deutsches Museum, Munich.)

Dmitri I. Mendeleyev, who took the brilliant leap of creating the periodic table of the known elements. (Courtesy of Zhores Alferov.)

Lothar Meyer, the chemist who pointed out that triads of elements, such as lithium, sodium, and potassium, and magnesium, calcium, and strontium, had similar chemical properties. His discovery was a first step toward the periodic table. (Courtesy of the Deutsches Museum, Munich.)

Robert A. Hadfield, a pioneer in the study of ferro-silicon alloys. He discovered their special magnetic properties, which were exploited further by E. Gumlich. (Courtesy of the Royal Society of London.)

René Just-Haüy, the mineralogist who discovered by careful measurement that the angles between cleavage planes produced in different specimens of the same mineral were identical. He concluded that the basic units of crystals are arranged in a regular three-dimensional lattice. He is regarded as the father of modern crystallography. (Courtesy of the Archives of the Academie des Sciences, Paris.)

August Bravais, who in 1850 demonstrated that the limitation on the number of macroscopic symmetry types found in nature to thirty-two can be related to the fact that the constituents form a regular three-dimensional lattice. This discovery induced others to explore the full range of symmetry, including translational symmetry, of such space lattices. (Courtesy of the Archives of the Academie des Sciences, Paris.)

Antoine H. Becquerel, who discovered
natural radioactivity incidental to his
study of the fluorescence of crystalline
materials. (Courtesy of the American
Institute of Physics Emilio
Segrè Visual Archives.)

Pierre Curie, who, with his brother,
discovered piezoelectricity in 1880
and later, with his wife, Marie, discovered radium. (Courtesy of the
American Institute of Physics
Meggers Gallery of Nobel Laureates.)

Woldemar Voigt, one of the leading solid-state physicists of his era. He published an encyclopedic summary of the research on the macroscopic physical properties of crystals carried out by him and others. (Courtesy of the Deutsches Museum, Munich.)

James Clerk Maxwell, who, among many other things, consolidated and extended the equations of electromagnetism. (Courtesy of the Science and Technology Museum and the Science and Society Picture Library, Kensington, London.)

Hermann von Helmholtz, one of the great generalist-scientists of the nineteenth century. He persuaded his brilliant young colleague Heinrich Hertz to investigate the possibility of generating Maxwell's electromagnetic waves with conventional circuit elements. He also encouraged Henry Rowland to test the validity of Maxwell's hypothesis that a rapidly moving electric charge carries a magnetic field with it. (Courtesy of the Deutsches Museum, Munich.)

Heinrich Hertz, who applied Maxwell's electromagnetic equations to the problem of generating electromagnetic waves on a laboratory scale. (Courtesy of the Deutsches Museum, Munich.)

Hertz's dipole oscillator-antenna. The two halves of the system were charged to opposite polarity until breakdown occurred in the gap between the small spheres. Electrical oscillation and associated electromagnetic radiation then occurred. (Courtesy of the Deutsches Museum, Munich.)

Henry A. Rowland, perhaps best known for the development of a machine for ruling fine optical gratings. He was
an expert in many areas of physics, exploratory and applied.
(Portrait by Thomas Eakins, 1897, oil on canvas, 80¼ × 54
in., 1931.5, Gift of Stephen C. Clark. © Addison Gallery of
American Art, Phillips Academy, Andover, Massachusetts.
All Rights Reserved.)

NOTES

1. The complexities of patent procedures as relevant to the invention of the integrated circuit are described in some detail in W. R. Runyan and K. E. Bean, *Semiconductor Integrated Circuit Processing Technology* (Menlo Park, Calif.: Addison-Wesley, 1990); see pp. 8ff.

2. A biography, as well as a personal account of Lavoisier's chemical research, can be found in the "Great Books of the Western World," *Encyclopedia Brittanica*, vol. 45 (1952). The Academie des Sciences has published a group of essays covering Lavoisier's many activities; see *Il y a 200 Ans Lavoisier* (Paris: Institut de France Press, 1994).

3. For a biography of Joseph Priestley, see John Graham Gillam, *The Crucible* (London: Robert Hale, 1954).

4. Robert Cahn of Cambridge University, Charles Wert and James Koehler of the University of Illinois at Urbana-Champaign, Leslie Reynolds of the Grainger Engineering Library of the same institution, and Robert M. Ehrenreich of the National Research Council have found no less than thirty-five references to the use of silicon in metallurgy in the period between 1888 and 1903, most clustered near 1900. We are indebted to them for this information. For information on ferrosilicon alloys, see Marion Howe, "The Metallurgy of Steel," *The Engineering Mining Journal* (London and New York, 1904); E. Greiner, J. Marsh, and B. Stoughton, *Alloys of Iron and Silicon* (New York: McGraw-Hill, 1933).

5. An account of the life and research of Robert A. Hadfield can be found in the *Obituary Notices of the Royal Society,* vol. 3, no. 8 (1940), p. 647.

6. See Woldemar Voigt, *Lehrbuch der Kristalphysik* (Leipzig: B. G. Teubner, 1928). This is a second edition, with additions by M. von Laue.

7. For an excellent biography of Ferdinand Braun, see Friedrich Kurylo and Charles Susskind, *Ferdinand Braun* (Cambridge, Mass.: MIT Press, 1981). Braun gave an account, in German, of his research up to 1909 in his Nobel lecture, available in the archives of the Nobel Foundation, Stockholm.

8. See the biography *Heinrich Hertz: Memoirs, Letters, Diaries,* 2d ed., prepared by Mathilde Hertz and Charles Susskind, arranged by Johanna Hertz (San Francisco: San Francisco Press, 1977). Also John H. Bryant, *Heinrich Hertz: The Beginning of Microwaves* (Piscataway, N.J.: IEEE Center for the History of Electronics, 1988); Charles Susskind, *Heinrich Hertz: A Short Life* (San Francisco: San Francisco Press, 1995).

9. A biography of Henry A. Rowland appears in *Men of the National Academy of Sciences,* vol. 5 (1912). See also *The Physical Papers of Henry Augustus Rowland* (Baltimore: Johns Hopkins University Press, 1902).

TWO

WIRELESS TELEGRAPHY

$\mathcal{G}$UGLIELMO MARCONI undoubtedly deserves the most credit for taking the first practical steps toward the age of wireless.[1] The young Marconi, without formal engineering education, became fascinated with Hertz's discovery of electromagnetic waves and remained driven by a desire to advance wireless technology for most of the remainder of his life. He also possessed the qualities of a persuasive entrepreneur. As early as 1895, when only twenty-one years old, he began experimenting with a reproduction of Hertz's equipment in his parents' home and garden in Italy in order to see how far electromagnetic waves could be sent and received, improving both transmission and reception, step by step, in the process.

Marconi's initial successes attracted the attention in Britain of the postal service and business leaders who were interested in linking the farflung components of the British Empire with better communications. As a result, Marconi received substantial financial support for his work from English sources, encouraging him to find a replacement for, or complement to, the long-distance, wire-bound, telegraphic system of the day. The new technology, it was hoped, would use the same form of coded language, but would avoid the need to extend lines of wire over long distances. As part of his program of telegraphy, Marconi developed devices that could send out a series of Hertzian waves. This made it possible to use the standard Morse code.

Once Marconi gained prominence, he had little difficulty in obtaining much technical advice, particularly from expert, admiring, and enthusiastic Italian engineers.

EDOUARD BRANLY: THE COHERER

The initial methods of detecting transmitted waves used by Marconi were based on a form of inductive receiver, initially one that generated a tiny arc, as used by Hertz, and which was adequate for laboratory experiments. Since the received signals were very weak, methods of amplification were sought. Fortunately the French scientist-inventor Edouard Branly had, in 1890, developed a device that proved to be very useful at the time.[2] He named the invention *tube de limaille* (tube with filings), but Oliver Lodge, who was enthusiastic about both it and Hertz's discoveries, gave the device the name "coherer," which stuck, at least in English. It consisted of a cylindrical tube with conducting leads at both ends, which contained a cluster of metal filings with appropriate contacts to the leads. The unit was placed in series as part of the resonant receiving circuit. The surge of current induced in that circuit served to align the metal filings in the direction of current flow and thereby increase the electrical conductivity of the coherer. The unit could simultaneously be part of a powered relay circuit that amplified the signal. In fact the amplified signal could even be used to print out the coded message on recording paper.

One of the great disadvantages of Branly's coherer was the need to produce decoherence of the filings after each pulse, a process accomplished by a mechanical blow to the tube, employing something in the nature of an electromechanical bell ringer. An ingenious alternative that did not require decoherence was developed by an Italian naval research group that worked closely with Marconi.[3] The two individuals most heavily involved were a signal corps officer, Paolo Castelli, who carried the rank of *Semaforista,* and Lieutenant Luigi Solari. Castelli seems to have had the primary concept. It involved the use of one or two mercury droplets sandwiched between two cylinders, one of iron and another of a material such as carbon. The mercury droplet dilated on passage of a pulse, changing the electrical resistivity of the combination. When it worked at its best, the mercury droplet device was self-restoring and did not require mechanical decoherence. It was, however, considerably less sensitive than Branly's coherer. It gained historical distinction in the first successful transatlantic experiments in 1901, in which Marconi received the letter "S" across the ocean.

The highly inventive Italian naval team also appears to have been the first to use telephonic earphones, presumably initially in an auxiliary circuit for "listening" to the pulses.

Fortunately, an entirely new form of reception was about to enter the picture, namely, that which involved rectification of electric currents induced by the incoming pulses.

RECTIFICATION

With the turn of the century, it became recognized that the very high frequencies of transmitted radiation were not very effective for inducing mechanical motion in useful devices such as magnetically activated earphones. As a result, there began a search for means of rectifying the currents generated by the received signal, converting them from alternating to direct form, and making their pulse duration characteristics more nearly compatible with the inertial properties of electromechanical equipment. One of the first successful systems for providing rectification was developed by R. A. Fessenden, a highly productive private inventor who employed a metal wire dipped in an electrolytic medium.[4]

An entirely new avenue, however, was opened by the brilliant Indian scientist J. C. Bose, who had carried out much research on the coherer and other instruments at his laboratory in Calcutta.[5] He applied for a U.S. patent for the use of galena (PbS) as a crystal rectifier in 1901 and was granted such a patent in 1904. This opened a search for other rectifying crystals that might have qualities comparable or superior to those of galena. The coherer was soon phased out. Direct listening of received signals with earphones became commonplace as a result of the relative convenience of a system powered by the energy of the incoming system.

FERDINAND BRAUN'S INVENTION: RESONANCE

Although Marconi had made remarkable progress in transmission and reception during experiments carried out over land in England by 1897, his system seemed to have attained what appeared to be a limited range of about fifteen kilometers. He faced difficulties and undoubtedly raised doubts in the minds of his sponsors. In the meantime Ferdinand Braun, discoverer in 1874 of crystal rectification, had risen to become head of a physics institute at the University of Strasbourg.[6] Among other things, he had, in 1897, invented the electron oscilloscope, originally known as the Braun tube. Inevitably, he became interested in the production and reception of electromagnetic waves. He prematurely sought ways of developing amplifying resonant circuits for wireless.

Finding the coherer very unsatisfactory in his early field work, Braun attempted to replace it directly with a crystal rectifier, with almost equally unsatisfactory results. The situation changed radically, however, when he began using earphones and gained the benefits of improved dynamic matching. Unlike Bose, he apparently did not attempt to patent this use of crystal rectifiers. Bose deserves special recognition for bringing public and professional attention to the merits of crystal rectifiers.

In the same eventful year, 1897, Braun was approached by a group of German entrepreneurs who asked him if he would cooperate in the development of a wireless program that might circumvent the monopolizing patents of the Marconi Company. He suggested that they investigate the possibility of transmitting waves through the volume or along the surface of water and soil and began experiments in canals. During the course of this research, he began to wonder about the limited range of transmission that Marconi was encountering in his field experimentation in England, since it did not seem to match his own experience and expectations. Then one day, quite by chance, he saw a photograph of one of Marconi's most advanced systems being used in field experiments and immediately spotted the source of the trouble. Marconi, unlike Braun, was not a highly trained scientist-engineer and had placed the oscillating circuit in direct line with the antenna and the ground lead. As a result, much of the energy that should have gone into transmission was dissipated in circuitry. Braun first tested and then, in 1898, patented a new concept, namely, maintaining the oscillating circuit essentially isolated except for coupling it inductively to the antenna and having the two in resonance with one another. This basic concept has been at the heart of wireless transmission and reception ever since.

Under normal circumstances, this patent could have reaped a fortune for Braun and his group of investors, but fate intervened. Just at that moment an American entrepreneurial group involved in the financial support of Nikola Tesla, inventor of an alternating current generator, the induction motor, and the so-called Tesla induction coil, began to make exaggerated claims concerning the ability of Tesla's actual and postulated inventions to transmit messages over long distances.[7] The international publicity given to these claims made it appear that other inventions might be obsolete. As a result, the lawyers involved with Braun's group delayed negotiating with competitors for nearly two years, by which time Braun's invention had been adopted without licensing arrangements by all groups, including that of Marconi, who was transmitting signals across the English Channel by 1899. Two years later, wireless telegraphy bridged the Atlantic Ocean and was receiving very serious attention from the international technical and industrial community. In the meantime, the size of equipment, including antennae, was growing to obtain greater outputs and ranges, with consequent decrease in transmitted frequency. Marconi, possibly without knowing it, eventually used wavelengths of the order of ten kilometers[8] (thirty kilohertz) in his transatlantic broadcasts.

There was much confusion initially within German governmental and industrial circles about how to concentrate the national effort in wireless telegraphy since three competing groups had entered the field and were working at cross purposes. In 1899, the three were joined to form the Telefunken Company. Braun, incidentally, despised the common use of the word *Funk* (arc) in connection with wireless telegraphy since the arc served primarily as a rapid switch in the circuit and was not the source of wireless waves. He felt certain that the arc would eventually be replaced when better oscillating circuits were developed. The word, however, had become too deeply ingrained in the popular vocabulary in Germany to be dropped and has been retained ever since. Even during World War II, personnel in the signal corps of the German army were referred to as *Funkers* (arcers). Deutsche Rundfunk is still the name of a television broadcasting network.

Although Braun had a deeper understanding of the fundamentals of electromagnetic theory in general, and of wireless in particular, than anyone else involved with Telefunken, his suggestions and inventions were often slighted by general management at considerable commercial cost. For example, he developed methods for increasing the transmitting power by having several capacitively charged sources feed into the primary oscillating circuit simultaneously. They were not used. His proposals to attempt to replace the coherer with a crystal rectifier were not well supported. He was, incidentally, probably the first individual to appreciate the fact that Marconi's ever-enlarging transmission system was emitting wavelengths much longer than those used by Hertz in the latter's laboratory equipment.

Braun lost considerable monetary reward through the failure of the German legal, business, and governmental communities to provide adequate support to back the rights merited by his patent. This was compensated in part by the action of the Nobel Prize Committee in 1909, which decided that his invention was fully as important as the work of Marconi and granted the Nobel Prize in physics to both that year. He rejected the opportunity to become a national celebrity in Germany with a high post in Berlin, and reverted to provincial life after receiving royal accolades.

Braun's life did not end happily. He journeyed to the United States by way of Norway at the end of 1914 to attend to business and legal affairs and became trapped in America, separated from family and colleagues, because of the shipping blockade of World War I. Eventually he was classified as an enemy alien. He died in Brooklyn, New York, on April 20, 1918, seven months before the end of the war. His wife had died in Germany in 1917.

ALEXANDER POPOV

The Russian physicist-engineer Alexander Popov, working in Russia, carried on research and development parallel to that described above at about the same time and with comparable initial success.[9] He developed very effective transmitters and receivers, starting as early as 1895, and was soon engaged in close cooperation with organizations in France and Germany, particularly the latter, forming a joint venture with what is now the Siemens Company. Practical equipment that operated over a range of about one hundred kilometers was rapidly put into service for operation on land and sea. There is no doubt, however, that the work of Marconi and Braun had much greater international impact as a result of its widespread application, mainly because of Marconi's entrepreneurial talents and the strong support for his patent monopoly that he received in England.[10]

Popov, incidentally, made the first careful scientific studies of electromagnetic radiation produced naturally in the atmosphere, that is, natural "static" signals, such as those produced by lightning and aurora. Unfortunately, he died of a stroke in his forty-sixth year.

THE KENNELLY-HEAVISIDE LAYERS

One of the very agreeable surprises experienced by Marconi and his colleagues while extending their work was the discovery that the curvature of the earth did not seem to limit the distances over which sufficiently long electromagnetic waves could be transmitted. The ionized layers of the upper atmosphere formed a reflecting mirror for an appropriate range of long wavelengths, above about fifteen meters by day and twenty by night, channeling such waves for travel over great distances and making transatlantic transmission possible. This discovery opened up, incidentally, opportunities for a great deal of research on the ionization layers themselves, sometimes called the Kennelly-Heaviside layers. The eventual development, in the 1920s and 1930s, of specialized equipment for studying the layers contributed much to the foundation of later work on radar.

LOW EFFICIENCY

The early form of wireless telegraphy based on the use of an arc-triggered transmitter, which was actually in use through World War I, was very inefficient. Not only were most of the frequencies emitted by the system well out of the range of normal hearing, even with sensitive earphones, but the transmitted beam was not strongly directional.

As a consequence of the low efficiency of wireless telegraphy at this stage, the U.S. Navy adopted circuitry involving the very powerful Poulson arc,[11] invented by the Dane Vladimir Poulson, for its transoceanic wireless communications during World War I. This arc system, which operated with the use of a confined gas and a magnetic field, had the great advantage that it produced a relatively uniform flow of pulses of electromagnetic waves, with an overall envelope approaching square-wave shape. The repetition rate of the pulses was in the audible range of frequencies.

One possible remedy for the frequency limitations of the human ear was proposed as early as 1902 by R. A. Fessenden, namely, what is now known as the heterodyne principle. He suggested that the most prominent frequencies in the transmitted signal be combined through a nonlinear element with a fixed "local" signal having a frequency that differs from the principal received frequencies by an audible increment. The audible undertone beat frequencies resulting from such "mixing" could be heard through earphones. Actually, use of the heterodyne principle was not to become really practical until reliable vacuum tubes became available, along with single-band transmission and stable local oscillators, a decade and a half later, that is, about the time of World War I.

Another method of enhancing the audio component of the transmitted signal from an arc was the introduction of a spinning notched wheel into the arc circuit in order to modulate the emission with an audible frequency and thereby increase the magnitude of that component. Somewhat the same result could be obtained by spinning an appropriately designed electrode of the arc.

DIRECTION FINDING

With rising interest in wireless, it was soon realized that the signals received from a given source could provide the basis for determining the direction of the source and finding its location by means of triangulation. In fact, E. Bellini and A. Tosi successfully demonstrated the process in 1906, developing a form of technology that was eventually used by the British Navy in World War I to locate the German fleet when it moved from its base in the North Sea in 1916 in readiness for what became known as the Battle of Jutland.

THE FLEMING VALVE: POWERED RECTIFICATION

In 1904, John A. Fleming introduced a new element into wireless that was to have far-reaching consequences, namely, the rectifying vacuum tube diode, or Fleming valve.[12] The device was based on Thomas A. Edison's observation in 1883 that the hot filament of an Edison lamp emitted a negatively charged current (electrons) that could be collected by a second electrode, commonly called the plate. The plate was attached to the positive terminal of the battery that fed heating power to the filament. The voltage drop in the filament was sufficient to provide it with a negative voltage relative to the plate. This system had been studied by several investigators whose main interest lay in trying to understand the laws governing such thermionic emission. For example, J. Elster and H. Geitel inserted a separate variable voltage source between the filament and plate to study the energy distribution of the emitted electrons; A. Wehnelt demonstrated that a platinum or tungsten filament coated with barium oxide had a more copious emission of electrons than a bare one at a given filament temperature.

Fleming realized that the system could be used to produce an amplified rectified current when an alternating voltage was applied between filament and plate and hence could convert the alternating currents received in the antenna into amplified direct current. This could be used, for example, to obtain relatively strong responses with magnetic earphones when all functioned well. For reasons that are not clearly understood, his patent, issued in 1904, focused on the original electrode arrangement developed by Edison, without the added voltage source introduced by Elster and Geitel. This made it possible for Lee De Forest to obtain an independent patent for the more flexible arrangement in 1905, leading inevitably to controversy.

In any event, Marconi and Fleming were soon cooperating in advancing the use of vacuum tube rectification. It should be emphasized again that no means for further, powered amplification of the rectified signal received through the earphones was available at the time of Fleming's invention. That would require a triode device.

ADVANCES IN CRYSTAL RECTIFIERS

The initial versions of Fleming's rectifying tube were expensive and far from perfect, being gassy since vacuum technology was still very primitive. Vacuum tubes really became reliable only after the development of the diffusion pump and the refinement of the composition of so-called chemical "getters," which could capture and retain residual gasses. In fact, the search for suitable chemical agents that could be inserted into light bulbs and other forms of vacuum tube, in order to control such gasses, became an important area of chemical research immediately after Edison's invention of the electric lamp. It has continued throughout the twentieth century.

Bose's earlier proposal of using crystal rectification of the type Braun had discovered in semiconductors came to the fore, and was soon very popular everywhere. Crystals of many different types and from many sources were tested for behavior and quality. The era of semiconductor research had begun on an extended scale under conditions in which it appeared, to a degree, as if the blind were leading the blind. To the practicing wireless engineer of the time, the class of materials that we now call semiconductors, a term (*Halbleiter*) apparently first adopted widely in Germany in the 1920s, appeared useful but bizarre. In addition to displaying the contact rectification first observed by Braun, semiconductors exhibited a type of internal electrical conductivity that was relatively feeble compared to that found in highly conducting metals such as copper. Also, unlike such metals, the electrical conductivity rose with increasing temperature. It was finally left to a special group of doughty physical chemists and physicists to bring a degree of order into what seemed to be a hopelessly chaotic situation (see chapters 4 and 5).

Very early in the work with crystal rectification, it was discovered that the best results were obtained with a unit in which one of the electrodes in contact with the semiconductor consisted of a relatively fine metallic wire, the so-called cat whisker. The opposite metallic contact had a broader area. Since most of the semiconducting crystals available at the time were either relatively impure natural minerals, such as galena (PbS), or crude, synthetic granular materials, such as silicon, silicon carbide, or graphite, all made for other purposes, the results obtained in testing units made of them were somewhat erratic. Both suppliers of such materials and wireless operators had their favorite specimens. The operators were accustomed to adjusting the contact site of the cat whisker in searches for what were called "hot spots." Such variability, bordering on what seemed the mystical, plagued the early history of crystal detectors, causing most vacuum tube experts of the post–World War I generation to regard them as being close to disreputable, and best forgotten. Fortunately this was not true for all such experts.

SILICON ENTERS THE TECHNOLOGY

It was soon noted that, with the customary trial and error in choice of specimen—and appropriate luck—it was possible to make relatively fine rectifying units using the metallurgical grade of crystalline silicon that had become available for industrial use and was pure to the level of a percent or so, at least as far as the elements that influenced electrical behavior are concerned. In fact, a patent concerning its use was issued to G. W. Pickard in 1906, a landmark year in this history.[13] A year later, H. C. C. Dunwoody obtained a similar patent for silicon carbide, then commonly used as an abrasive.[14]

Pickard was born in 1877 and attended both Harvard University and the Massachusetts Institute of Technology. He became actively interested in wireless telegraphy at the turn of the century and accepted a research position at the American Telephone and Telegraph Company, where he was employed between 1902 and 1906. During this period he became much interested in the use of crystal rectifiers for reception and explored many combinations of metal semiconductor junctions. Commercial-grade silicon obtained from the Westinghouse Electric Company proved to be the best. He obtained a patent on the use of silicon in rectifying diodes during his years at AT&T. Then, in 1907, he and two associates organized a company to market his patented detectors. It appears that Pickard took the matter of crystal rectifiers very seriously indeed. In the course of his research, he tested over thirty thousand combinations of materials.

To summarize, crystal rectifiers, including units involving silicon, played a major role in the evolution of coded wireless detection almost from the start of the applied technology and were in prominent use through World War I. A pamphlet that reviewed the design and use of crystal rectifiers in 1918 was issued by the U.S. Army Signal Corps just after the war ended. Among other types, silicon combined with a steel cat-whisker was given a prominent position in the pamphlet, indicating that its service had been highly regarded. As might be expected, the growing use and importance of vacuum tubes, which were in a rapid state of improvement and development at that time, was also emphasized in the document.

Although the use of cat-whisker diodes for mainstream practical work essentially vanished soon after World War I when vacuum tube circuitry became commonplace, crystal rectifiers did not disappear completely. Large area rectifiers of both selenium and cuprous oxide found use as "power" converters from alternating to direct current since they were inexpensive, durable, and did not require an independent power source for their operation. Stacked discs of silicon carbide, incidentally, found special use as a series link in grounded lightning arrestors for power lines. Normally the material exhibited relatively low electrical conductivity, but a form of "avalanche" electronic breakdown was induced in it when a high voltage was applied. The breakdown would, however, heal itself after the voltage pulse induced by the lightning ended, returning the arrestor essentially to its initial state.

COMMERCIAL BROADCASTING

The first commercial broadcasting station in the United States, KDKA, was opened in Pittsburgh in 1920 by the Westinghouse Electric Corporation, and was soon followed by others throughout the country. Amateur radio sets constructed by impecunious teen-agers in the early 1920s contained cat-whisker rectifiers, usually made with galena rather than silicon. As in the days of crystal-rectified wireless telegraphy, these amateur sets drew their power directly from the received signal of the antenna so that the range of reception from modestly powered radio stations was relatively small. At that time, a single triode vacuum tube purchased in the open market at one of the newly opened "radio stores" where the amateurs congregated cost five dollars and was highly prized as a Christmas present.

It may be added that one of the great blessings of that period for both amateur and professional radio enthusiasts, derived from recent advances in chemistry, was the ready availability of inexpensive panels of Bakelite, a black synthetic resin invented in 1909 by L. H. Baekeland. It was a hard-setting, electrically insulating polymer formed by combining a phenol material with formaldehyde. It served well both as a mounting board and as a dark, mirrorlike face-plate on which the numerous tuning knobs, common at the time, were located.

Guglielmo Marconi, who exploited Hertz's method of generating electromagnetic waves for commercial purposes. He was not technically trained but
had great sense of purpose and was an excellent entrepreneur. He is responsible
for the speed with which wireless telegraphy came into international service.
(Courtesy of the Marconi Company Archives.)

Edouard Branly,
the inventor of
the coherer and
a distinguished
pioneer in the
evolution of
wireless telegra-
phy. (Courtesy
of Musée
Branly, Paris.)

R. A. Fessenden *(seated)*, a turn-of-the-century inventor, with his staff. He invented one of the first useful rectifiers for wireless based on the use of a wire immersed in an electrolytic medium. He also proposed using the heterodyne principle for enhancing the audio reception of wireless signals. (Courtesy of the Smithsonian Institution.)

J. C. Bose, the Indian inventor who apparently was the first individual to employ a crystal rectifier in wireless telegraphy in a truly practical way. He successfully applied for a patent for the use of galena, lead sulfide, for this purpose in 1901. (Courtesy of the Royal Society of London.)

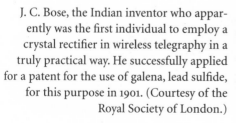

Ferdinand Braun, who shared the
Nobel Prize with Marconi in 1909.
He discovered crystal rectification
in 1874, invented the electron oscil-
lograph (cathode ray tube), and
developed the proper resonant relationships between the primary
oscillator and the antenna to opti-
mize wireless transmission, for
which he was awarded the prize.
(Courtesy of the Tuesday Morning
Club of Tübingen, the Library of
the University of Tübingen,
Professor H. Metzner, and
Mr. Fritz Seitz.)

Nikola Tesla, whose many inventions included a practical means of
converting alternating and direct
current into one another, the induction motor, and the Tesla induction coil. (Courtesy of Mark
Rothenberg and the Smithsonian
Institution.)

Alexander Popov, the Russian scientist-inventor who successfully carried out research on wireless telegraphy in parallel with Marconi and Braun. (Courtesy of Academician Zhores Alferov.)

John A. Fleming, who applied Thomas A. Edison's discovery that a heated filament in an evacuated bulb emits electrons that can be collected by a neighboring electrode to rectify wireless signals. The so-called Fleming valve was patented in 1903. (Courtesy of the Royal Society of London.)

(Above) Thomas A. Edison, the prolific American inventor who discovered that the hot filament of a light bulb emits negative charges, which eventually were demonstrated to be electrons. This phenomenon was known initially as the Edison effect. He is shown here at the time of a meeting of the National Academy of Sciences in 1878 where he gave a demonstration of an experimental version of the phonograph. (Courtesy of the U.S. Department of Interior, National Park Service, Edison National Historic Site.)

(Right) G. W. Pickard, who discovered the excellent rectifying properties of silicon crystals in 1905, using metallurgical grade material—a historic milestone. (Courtesy of the Center for the History of Electrical Engineering of the Institute of Electrical and Electronics Engineers, Rutgers–The State University of New Jersey.)

Two crystal rectifier units manufactured for use in wireless telegraphy. They were made available soon after the developmental work of Bose, Pickard, and Dunwoody. The silicon diode on the right was produced by Pickard's company. (Courtesy of Mark Rothenberg and the Smithsonian Institution.)

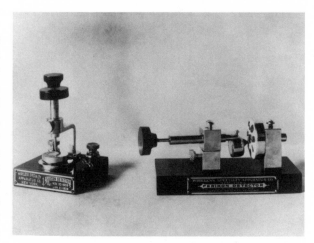

The Principles Underlying Radio Communication

▽

RADIO PAMPHLET No. 40
December 10, 1918

———

Signal Corps, U. S. Army

The cover page of a technical pamphlet on the status of military wireless in World War I. It was issued by the Signal Corps of the U.S. Army in 1918 at the end of that war. Silicon crystal rectifiers are featured, although it is recognized that the vacuum tube era is at hand. (Courtesy of Jack S. Kilby.)

Washington : Government Printing Office : 1919

NOTES

1. Two relatively recent books on Marconi are W. P. Jolly, *Marconi* (New York: Stein and Day, 1972); Orin E. Dunlap, *Marconi: The Man and His Wireless* (New York: Arno Press, 1991). *Marconi: Whisper in the Air* (Archer Films Limited, 1994; marketed by New Video Group, New York) is a one-hour video documentary occasionally shown on public television. See also W. J. Baker, *A History of the Marconi Company* (London: Methuen, 1974); Probir K. B. Bondyopadhyay, "Guglielmo Marconi, the Father of Long Distance Radio Communication: An Engineer's Tribute," *Conference Proceedings, Twenty-fifth European Microwave Conference,* vol. 2 (1995), p. 879. The latter paper focuses on the contributions of several of the great pioneers. Bondyopadhyay had a close relationship with the Marconi family.

2. For a biography of Branly, see Philipe Monod-Broca, *Branly* (Paris: Belin Press, 1990).

3. See, for example, Augusto Righi and Bernardo Dessau, *La Telegrafia senza filo* (Bologna: Ditta Nicola Zanichelli, 1908), esp. pp. 367–68; Luigi Solari, *Storia della radio* (Milan: S. A. Fratelli Treves Editori, 1939), esp. pp. 28ff., 214ff. Admiral Ernesto Simion of the Italian Navy wrote an official history of the Italian research on wireless, *Il Contributo dato dalla R. Marina allo sviluppo della radiografia* (Rome: Ministry of the Navy, 1927). We are deeply indebted to Giuseppe Baldacchini and Gianfranco Chiarotti for providing copies of material from these books. We are also indebted to Dr. Probir K. B. Bondyopadhyay for the privilege of reading an unpublished article written by V. J. Phillips of the University of Wales, Swansea, dealing with this and related topics associated with the detection of the 1901 transatlantic wireless signal. The article was written for Dr. Bondyopadhyay early in 1991. In a paper by G. C. Corazza, "Marconi and the Invention of Wireless Communications," *Rendiconti Accademia Nazionale dei XL* 19 (1995): 77, mention is made of the use of earphones in the transatlantic tests. We are indebted to Franco Bassani for a copy of this paper.

4. See Bondyopadhyay, "Guglielmo Marconi."

5. J. C. Bose, U.S. Patent No. 755,840, 1904. Dr. Probir K. B. Bondyopadhyay informs me that Bose's application for a U.S. patent was made on Sept. 30, 1901.

6. For a biography of Braun, see Friedrich Kurylo and Charles Susskind, *Ferdinand Braun* (Cambridge, Mass.: MIT Press, 1981).

7. For a recent biography of Nikola Tesla, see Marc. J. Seifer, *Wizard: The Life and Times of Nikola Tesla* (New York: Birch Lane Press, 1996); see also, Margaret Cheney, *Tesla, Man Out of Time* (Englewood Cliffs, N.J.: Prentice-Hall, 1981). A French translation of the latter is also available; see *Tesla, la passion d'inventer* (Paris: Belin, 1987). An older biography is John J. O'Neill, *Prodigal Genius* (New York: Ives, Washburn, 1944). With regard to Tesla's New York financial supporters, Mark Twain commented at that time that many New York entrepreneurs were guided by five rules: Get rich, get rich quickly, get very rich, preferably by dishonest means, but by honest ones if no alternative is offered.

8. The relationship between the wave length λ and the frequency ν of electromagnetic radiation in a vacuum is $\lambda\nu = c$, where c is the velocity of light, namely 3.0×10^{10} cm per second.

9. For a technical biography of Popov, see Charles Susskind, *Popov and the Beginnings of Radiotelegraphy* (San Francisco: San Francisco Press, 1962). It seems reasonable to suppose that Popov, in developing his cooperative program with what is now the Siemens Company, did so in association with Nikolaus Riehl's father, who was in charge of the company's interests in Russia from the 1890s until 1918. See N. Riehl and F. Seitz, *Stalin's Captive: Nikolaus Riehl and the Soviet Race for the Bomb* (Washington, D.C.: American Chemical Society and the Chemical Heritage Foundation, 1996).

10. An interesting discussion of personal priorities for the development of wireless transmission of information is offered in the article "Marconi a-t-il inventé la radio?" by the French science historian Jean Cazenobe, in the semipopular scientific journal *La Recherche* 26 (1995): 508. He concludes that credit for the invention is to be broadly distributed, although Marconi was an individual with an especially intense vision. Marconi's own declarations over the years were relatively modest, although he stated that he trusts it is clear that he was "there."

11. An account of the Poulson arc can be found in A. Williams, *How It Works*, 11th ed. (London: Thomas Nelson, 1922).

12. Fleming has written an autobiography: *Memories of A Scientific Life* (London: Marshall, Morgan and Scott, 1934). Additional biographical material by W. H. Eccles appears in the *Obituary Notices of the Royal Society*, vol. 5, no. 4 (1945), p. 231.

13. G. W. Pickard, Means of Receiving Intelligence Communicated by Electric Waves, U.S. Patent No. 836,531, 1906. See James E. Brittain, "Greenleaf W. Pickard and the Eclipse Network," *Proceedings of the IEEE* 83 (1995): 1434.

14. The current status of development of silicon carbide as a useful semiconductor is reviewed in several papers in the Materials Research Society bulletin: see *Silicon Carbide Electronic Materials and Devices* 22, no. 3 (1987). Such devices can play an important role at high temperatures and high frequencies, where silicon becomes ineffective. A field-effect transistor operating at 40 gigahertz has been demonstrated. The lattice structure exhibits several polytypes. The (indirect) band gap ranges between about 2.4 eV for the cubic zincblende structure to 3.35 eV for the hexagonal wurtzite structure. See also the collection of articles on silicon carbide in *Physica status solidi (a)*, Applied Research 1, no. 1 (July 1997).

THREE

VACUUM TUBE ERA

*I*n 1906, Lee De Forest's interest in vacuum tube diodes led him to realize that it might be possible to control or modulate the flow of electrons from filament to anode in the diode by inserting a third electrode with its own independently variable voltage into the tube.[1] He patented the device and named it the *audion,* somewhat to the distress of John Fleming, who had hoped that his name would be attached to future electron tubes. De Forest obtained a second patent a year later in which the third electrode was designed in the form of a two-dimensional wire mesh, or "grid," through which electrons could pass, thus giving a specific name to the third electrode.

De Forest's choice of name for his three-electrode tube shows that he had some conception of its ultimate potential for amplifying and reproducing sound. It is not unreasonable to assume that he developed a primitive amplifier and oscillator, as he later claimed. He did not, however, have access to the research facilities of a large corporation nor did he have Marconi's entrepreneurial gifts to develop his own. It was left to others to make fuller use of his invention. In any event, the device was initially considered to be somewhat exotic; it was little understood and saw little use at the practical level other than as a diode. Moreover, the initial version was very erratic for the same reason as the Fleming valve—it usually contained residual gas, which complicated its operation. Vacuum technology was very primitive in the early part of the twentieth century.[2]

The first individual to take such a triode seriously in a practical sense, as the essential element in an oscillating, feedback circuit, was the Austrian engineer Alexander Meissner, who provided a demonstration of such a circuit in 1910. It appears that this observation led to occasional use of the tube in place of an arc-triggered oscillating primary circuit in wireless transmission. In a crude sense, such use represented a forerunner of the development of radio; however, the device was still much too remote from practical development to offer more than the basis for speculation in 1910.

American Telephone and Telegraph (AT&T) purchased the rights to the audion from De Forest in 1912 as part of a broad, intense search for technology that would permit it to extend the telephone system throughout the United States (see chapter 13). The company was committed to this task by its head, Theodore Vail, and it achieved success through the use of a much improved triode in 1915.[3] In the meantime, the General Electric Company, with the brilliant help of Irving Langmuir, a physical chemist, had also developed a reasonably perfected triode, along with graduated improvements in vacuum technology to which Langmuir was the major contributor.

E. H. ARMSTRONG

In 1912, a remarkable twenty-two-year-old American amateur, E. H. Armstrong, who was fascinated by wireless technology, also discovered the great potential of positive feedback.[4] With improved vacuum tubes, he developed a reasonably stable regenerative oscillating circuit, thereby creating a new flexible and reliable source of narrow-band electromagnetic waves—the first since the work of Hertz. The triggering arc was finally, although gradually, to be replaced, in keeping with Ferdinand Braun's expectations. Significantly, Armstrong had the entrepreneurial drive to find supporters to push the new technology ahead.

It was now possible to use the heterodyne principle at the receiver end by employing a fixed-frequency oscillator and an appropriate mixer, in accordance with Fessenden's proposal a decade earlier. Moreover, the ingredients for efficient wireless transmission of voice messages were now available, as was the means to develop antenna arrays that would permit relatively precise directional transmission. In other words, the age of modern radio broadcasting was dawning, although extensive commercial use in the United States was to await the end of World War I.

Unfortunately, Armstrong and De Forest entered into a long and wasteful series of lawsuits over matters of priority concerning the invention of the regenerative circuit. Whereas De Forest would have been willing to compromise at one point, Armstrong insisted on continuing the struggle. In the end, the latter lost in court even though the engineering community has supported his claims of priority.

THE SUPERHETERODYNE CIRCUIT

Simultaneous invention of devices, circuits, or systems in different countries is common when independent groups are pursuing similar goals. Thus, in 1918 both Armstrong, in the United States, and Walter Schottky, in what was then wartime Germany, filed patents for superheterodyne circuitry.[5] This permitted several stages of amplification of a received radio signal to take place at inaudibly high frequencies prior to the final conversion of the received signal to the audio range. The advance in concept permitted greater compactness, with increased amplification, in radio receivers.

It may be noted in passing that the stabilization of emission frequencies was greatly assisted by the development of quartz crystal oscillators based on the discovery of the piezoelectric effect by the Curie brothers many decades earlier. Segments of crystals cut along axes in which the temperature dependence of the effect was near a minimum were preferred in practical frequency stabilizers.

EXTENSION OF FREQUENCIES: MICROWAVES

By the end of the 1920s, almost every middle-income home in the United States possessed a commercially produced radio, fashioned to the financial means and taste of the owners, and operating in the higher range of frequencies, eventually up to about 1.5 megahertz. Amateur radio transmission and reception had become popular much earlier in the decade, soon after World War I. To avoid conflict with commercial broadcasting, the amateurs worked at a higher range of frequencies, initially near and above 1.3 megahertz. Among other things, enthusiastic amateurs soon succeeded in demonstrating the feasibility of transmitting signals in these higher ranges over long distances. The Radio Club of America, an amateur organization, carried out the first "official" bridging of the Atlantic with a 1.35-megahertz transmitter on December 11, 1921. That success encouraged commercial research and development that eventually made the region up to about 100-megahertz accessible. Marconi's company shifted to shorter wavelengths, that is, higher frequencies, for long-distance transmission as a result of these new discoveries.

In fact, H. Barkhausen and K. Kurz in Germany were experimenting as early as 1920 with a tube that produced radiation in the one-meter—that is, in the hundred-megahertz—range, although somewhat erratically.[6] Electromagnetic radiation in this general range of frequencies was subsequently given the designation *microwaves*. The tube possessed a positively charged screenlike grid about which the electrons could oscillate in clusters, generating waves as a result. The anode, or *plate,* was negatively charged relative to the grid, creating a potential trough between the cathode and anode centered about the grid. Apparently the oscillations produced in the tube used by Barkhausen and Kurz were discovered somewhat accidentally. This occurred when they were testing the level of vacuum within the tube by adjusting the voltage on the anode to negative values in order to collect the positive ions present, and thereby determine their number, as a measure of the residual gas present.

THE MAGNETRON

Another such system was the magnetron, first proposed by the Swiss physicist H. Greinacher in 1912 as a means of determining the ratio of the charge to mass of the electron, but first produced as a precision device in 1921 by Albert W. Hull at the General Electric Research Laboratory.[7] It was left to others, however, to discover that the magnetron could serve as a source of microwaves. The leader in this development appears to have been E. Habann, who worked with a version in which the anode was split into semicylindrical halves.[8] In addition, he and others soon developed multi-segmented versions having a pattern of interconnecting wires (so-called *straps*) to control the frequencies at which the tubes would oscillate. Such devices were the predecessors of what eventually developed into a very powerful source of microwave radiation for use in radar.[9] It was one of the favorite sources of such radiation in the various experiments carried out in the microwave range by Hans E. Hollmann (see chapter 7).[10]

FREQUENCY MODULATION

Armstrong sought to take special advantage of the promising new developments in vacuum tube technology, and the higher range of frequencies that had become available in the 1920s, by commercializing an invention that he patented in 1933, namely, frequency modulated (FM) transmission and reception. This process greatly reduced the reception of bothersome, so-called "static," signals arising from various electromagnetic disturbances created either by natural or human activity, such as lightning or aurora, or spark discharges in electrical equipment. In this system, the deviation of the frequency of the received signal at each instant from that of the main carrier of the band determines the magnitude of the audio signal sent out from the terminal audio circuit. Thus the influence of extraneous, static frequencies does not emerge as sonic frequencies in the audio output. Apparently Hans Hollmann, mentioned above, developed a working version of the same system at the same time, but he did not receive a broadcasting license. The technology lingered dormant in Europe until after the war.

Incidentally, the commercial broadcasters of the period, who relied on more direct amplitude modulation (AM) of the carrier signal, tended to mute the higher end of the audio spectrum in order to reduce the contribution from static in that range, where it was particularly bothersome. As a result the reproduced sound of a coloratura soprano or a piccolo left much to be desired.

In any event, Armstrong, in the face of strong opposition from the well-established broadcasting companies in the United States, which were beginning to look forward to the development and commercialization of television and felt that FM radio was an expensive diversion, established his own FM transmission station in the New York City area, operating in the range from 42 to 50 megahertz, to demonstrate the feasibility of the new technology. It proved, however, to be a limited commercial success at that time since equipment for reception was not widely available and was expensive. He eventually took his own life in a state of depression brought on by a variety of personal problems.

FM radio, greatly preferred for audio reception, now uses a portion of the spectrum between 88 and 109 megahertz. It is worth noting in this connection that when commercial television was finally licensed after World War II, the Federal Communications Commission (FCC) in the United States insisted that the sonic component of the transmitted signal employ frequency modulation.[11]

TELEVISION IN THE 1930S

Experimental and commercial television developments were pushed ahead much faster in Europe in the 1930s than in the United States. The Federal Communications Commission feared that the available systems, mainly the one in development by the Radio Corporation of America under the leadership of Vladimir Zworykin, were not yet ready to be commercialized in a way that would give adequate benefit to the public for its investment in receivers. Indeed the actual feasibility of television was being demonstrated throughout much of Western Europe in the mid-1930s using a portion of the range of frequencies below that now reserved for the FM band. For example, England had two hours per day of experimental television broadcasting during the latter half of the 1930s. It was carried out by the British Broadcasting Company under government sponsorship. Denis M. Robinson, about whom much more will be related in connection with radar in chapter 10, started his professional career in England as one of the pioneers in the development of television. When World War II started, the British government terminated television broadcasting abruptly as part of the blackout and expanded its attention on radar.

Lee De Forest, the inventor of the triode vacuum tube, which he designated "the audion." Actually the tube saw little use until the next decade when refinements in vacuum technology and better theoretical understanding of the operation of the tube were achieved by AT&T and the General Electric Company. Noteworthy advances were also made by the Europeans. (Courtesy of the Center for the History of Electrical Engineering of the Institute of Electrical and Electronics Engineers, Rutgers–The State University of New Jersey.)

E. H. Armstrong, the brilliant inventor. He independently discovered how to use the triode vacuum tube for audio amplification, and thereby opened the door to modern radio transmission, at least in the United States. He also invented the regenerative circuit, the superheterodyne principle of amplification, and frequency modulated transmission of radio signals. (Courtesy of the Smithsonian Institution.)

Walter Schottky, who independently invented the superheterodyne principle and later developed one of the basic theories of crystal rectification. (Courtesy of the Deutsches Museum, Munich.)

Albert W. Hull, who, in 1921, perfected H. Greinacher's experimental device, the magnetron, for determining the ratio of the electronic charge and mass. (Courtesy of the National Academy of Sciences.)

Vladimir Zworykin *(left)*, the inventor of the iconoscope and the prime mover in the development of television in the United States. He is seen here with Manfred von Ardenne at an electronics exhibition. (Courtesy of the American Institute of Physics Emilio Segrè Visual Archives.)

NOTES

1. An excellent one-hour video documentary, *The Empire of the Air—The Men Who Made Radio* (Florentine Films, 1991), directed by Ken Burns, contains an account of the contributions of both De Forest and E. H Armstrong, as well as David Sarnoff, the entrepreneur.

2. An account of the history of vacuum technology and much of its advanced developments can be found in Saul Dushman, *Scientific Foundations of Vacuum Technique* (New York: Wiley, 1949). It appears that the Europeans were well aware of the potentialities of the triode vacuum tube and were also working very hard to improve vacuum technology. See, for example, the activities of W. Gaede described in Dushman's book.

3. For a biography of Theodore Vail, see Albert Bigelow Paine, *In One Man's Life* (New York: Harper, 1921). We are indebted to Mrs. Katherine M. Hurd, Vail's granddaughter, for access to this volume. In the twenty-year interval between Vail's two periods of employment with AT&T, he was, among other things, active in the electrification of communities in South America.

4. Information about E. H. Armstrong is conveyed in the documentary film *Empire of the Air*. For biographies of Armstrong, see James Brittain, "Edwin Howard Armstrong: An Independent Inventor in a Corporate Age," *Proceedings of the Radio Club of America* (1984), p. 121; Lawrence Lessing, *Man of High Fidelity: Edwin Howard Armstrong* (Philadelphia: Lippincott, 1956).

5. A technically oriented biography of Walter Schottky appears in an article (in German) by F. Paschke in *Siemens Forschung und Entwicklung-Berichte* 15 (Berlin: Springer, 1986), p. 287.

6. For additional information about the work of Barkhausen and Kurz, see H. E. Hollmann, *Physik und Technik der Ultrakurzen Wellen*, 2 vols. (Berlin: Springer, 1936).

7. See V. H. Greinacher, *Verhandlung der Deutscher Physik. Gesellschaft* 14 (1912): 856; A. W. Hull, *Physical Review* 18 (1921): 34, and 25 (1925): 645. A biography of Hull appears in the *Biographical Memoirs of the National Academy of Sciences*, vol. 41 (1970), p. 215.

8. See E. Habann, *Zeitschrift für Hochfrequenztechnik* 24 (1924): 115.

9. See Ulrich Kern, "Die Enstehung des Radar Verfahrens: Zur Geschichte der Radar Technik bis 1945" (Thesis, University of Stuttgart, 1984). The authors are indebted to Erich Hahne of the University of Stuttgart for a copy of this thesis. See also Henry E. Guerlac, *Radar in World War II*, 2 vols. (New York: American Institute of Physics, 1987) and idem, "Radio Background of Radar," *Journal of the Franklin Institute* 250 (1950): 285.

10. A brief biography of H. E. Hollmann was prepared by H. Frühauf for Hollmann's sixtieth birthday. It appears in the discontinued journal *Hochfrequenztechnik und Elektroakustik* 68, no. 5 (1959): 141. An English translation has been deposited at the Center for the History of Electrical Engineering of the Insti-

tute of Electrical and Electronics Engineers Incorporated and Rutgers–The State University of New Jersey.

11. In the period immediately after World War II, the victorious Allies placed the standard broadcasting stations in Germany—which employed amplitude modulation, like those in the United States—under very strict control. As a result, a number of stations developed auxiliary units that broadcast frequency modulated programs in the higher frequency bands, which were not under similar regulation. These steps accelerated the development and use of FM technology for radio and television broadcasting everywhere.

FOUR

SEMICONDUCTORS

$\mathscr{A}$LTHOUGH THE EARLY WIRELESS engineers and technicians found crystal rectifiers relatively inexpensive and reasonably convenient to use (in spite of their apparently idiosyncratic behavior), the underlying explanation for their operation and variability remained much of a mystery until a great deal of substantial research had been carried out by physical chemists and physicists. Indeed, full explanations could not be expected until the quantum theory of electron behavior had been worked out in some detail.[1] Moreover, anything representing complete clarification of the basic phenomenon of rectification extended over a number of decades, essentially well into the post–World War II period.

BUSCH'S HISTORICAL SURVEY

Georg Busch of the Swiss Federal Institute in Zurich (ETHZ), who has spent many years working with semiconductors, has carried out a valuable historical survey of research on them that adds a great deal to our general understanding.[2] His survey, with additions, is worth summarizing here.

In the period around 1700 when there was much interest in electrostatically charged bodies, it was discovered that such charge could be drawn off or conducted away by making contact with certain solids. Good conductors such as copper, silver, and gold were best, but there were also other substances that, while effective, acted much more slowly. Apparently Alessandro Volta, who was responsible for the first primitive form of the chemical electric battery, or "pile," late in the 1700s, made a study of the slow-acting materials that were neither good conductors nor good insulators and gave them the designation (in translation) "materials of semiconducting nature." Although the present-day use of the word "semiconductor" is of more recent origin, dating from the period around World War I, there was at least recognition of the fact that there are electrical conductors that are not metals.

DAVY AND FARADAY

Humphrey Davy, who began the study of electrochemistry in the early decades of the nineteenth century, immediately following Galvani's invention of the electric battery, investigated the conducting properties of many metals and concluded that their conductivity decreased with increasing temperature—a behavior that he decided was a general property of conducting solids. Following this, his brilliantly productive younger colleague Michael Faraday extended Davy's measurements to many compounds that were not metals and concluded that the opposite rule usually held in such cases, that is, they tended to become better electrical conductors at higher temperatures. Among many other things, he found that the electric conductivity of silver sulfide (Ag_2S) not only displayed this trend, but jumped to a level almost comparable to that of a metal at $175°$ C—a transition later demonstrated to be associated with a change in phase, or crystal structure, to the metallic state, to use modern terminology.

Faraday's studies of the electrical conductivity of relatively poor conductors were extended by the electrochemist Johann W. Hittdorf in the 1850s. Busch has reanalyzed Hittdorf's data for silver sulfide and cuprous sulfide (Cu_2S) that appeared in a paper published in 1851. In the first case, the measurements exhibited the discontinuity observed by Faraday, whereas those for the second were found to lie on a straight line when the logarithm of the conductivity was plotted as a function of the reciprocal of the absolute temperature, that is, in the representation known as the Boltzmann, or Arrhenius, plot. (The activation energy for silver sulfide, as represented by the slope of this line, turns out to be about 0.38 eV, or 8.74 kcal.) Being an experienced electrochemist, and having no good reason to think otherwise at that period, Hittdorf assumed that the conductivity is electrolytic, that is, it depended upon the migration of charged atoms, as in the solutions that he studied.

THE FOUR-PROBE METHOD

The true internal conductivity of a homogeneous semiconductor can be determined with the use of the four-probe arrangement shown schematically on page 65. The two outer probes are used to produce internal current flow with an applied voltage; the inner two probes are connected to a potentiometer that is varied so that no current flows in that auxiliary circuit. The internal conductivity can then be calculated from knowledge of the current between the outer probes, the balancing potential for the two inner probes, and the geometry. This experimental arrangement eliminates effects arising from blocking layers or other special sources of resistance at the contacts.

THE HALL EFFECT

H. A. Rowland wondered if the electrical current in a rectangular metal sheet would be deflected transversely by a magnetic field that is oriented normal to the plane of the sheet, in contradiction to a prediction made by Maxwell. When he failed to find such a transverse effect, he encouraged one of his advanced students, Edwin H. Hall, to pursue the topic using a more sophisticated design for the experiment, namely, searching for a potential difference transverse to the current and the magnetic field.[3] The effect was finally found by Hall in 1879 in gold leaf. While the observations supported the notion that the carriers of current were negative, apparently in keeping with the discovery of the negatively charged electron by J. J. Thomson in 1897, some metals, such as aluminum, displayed the opposite sign. It was as if the carriers were positively charged. Such observations of apparently positive carriers of current were subsequently designated "anomalous"—a mystery at the time since it was known that electrons, which were presumably responsible for the current, were always negatively charged. An explanation will be given later in this chapter.

Measurements of the Hall effect in semiconductors have provided a very valuable tool for determining two of the important parameters concerning the current carriers. When combined with the electrical conductivity, the Hall measurement permits determination of both the volume density of free carriers, assuming they are all of one sign, and their mobility, that is, their velocity of migration in the direction of a unit applied electric field.

THE FREE ELECTRON THEORY OF METALS

Following Thomson's discovery of the electron, Carl V. E. Riecke at the University of Leipzig carried out a year-long experiment at the turn of the century to see if an electrolytic effect could be observed in a copper bar placed in series with the power line serving the laboratory. Finding none, he concluded that the flow of current in metals having a high electrical conductivity was purely electronic in nature. This in turn led him and Paul Drude to propose that the conduction electrons in metals such as copper, silver, and gold exist in a form that might loosely be called a "free electron gas." The concept appeared to be very attractive at the time, since it was hoped that it might account for the high thermal as well as high electrical conductivity of such metals. Unfortunately, an attempt by Drude to develop a quantitative theory of the properties of such a gas led to highly contradictory results. For example, there was no evidence that the actual contribution of the electrons to the molar heat was 3R/2, where R is the gas constant, as one might have expected. The theory was to be revived in the late 1920s after the quantum statistical properties of electrons became more clearly understood.

THE ACTIVATION ENERGY

Johan Königsberger, one of the distinguished group of physicists who studied the chemical and physical properties of minerals, became aware of the problems that were emerging with respect to the origin of electrical properties of metals and semiconductors and offered his own contribution. He proposed, in effect, that the conduction electrons would be bound at the absolute zero of temperature, but would become increasingly free with rising temperatures, forming something akin to the gaseous vapor over a liquid or solid, with a finite heat of evaporation or sublimation. While this theory made little sense when applied to normal, high-conducting metals, whose electrical conductivity increases with decreasing temperature at low temperatures, it at least described in a qualitative way the general behavior of the conductivity of semiconductors, in which the carriers do need an activation energy to become free to form an electron gas in the sense of Drude and Riecke. Finally, something in the way of order was beginning to emerge. Apparently the name *semiconductor* (*Halbleiter*) first found common use in Königberger's laboratory.

KARL BAEDEKER

Between 1909 and 1911, Karl Baedeker took another great step forward in clarifying the basic properties of semiconductors.[4] He was, incidentally, the son of the individual who created the classic series of worldwide travel guides. The younger Baedeker used the best chemical and physical techniques available at the period to produce thin films of pure metals on glass or mica. These were then transformed into compounds, such as oxides, sulfides, and halides by exposure to appropriate vapors. The steps used were highly sophisticated for the time, at least in the field of semiconductor research. They represented a significant step forward in gaining control of chemical composition while working with a simple physical configuration. One is reminded somewhat, although in a very rudimentary way, of the techniques used to fabricate silicon chips today.

As his work on a variety of materials progressed, Baedeker discovered an illuminating effect in cuprous iodide (CuI), which he studied in detail. If permitted to come to stoichiometric equilibrium in air, the resistivity of the film was fairly high. When, however, the film was subsequently exposed to iodine vapor, the stoichiometric balance was upset by the absorption of additional iodine, which converted some of the monovalent copper atoms distributed throughout the volume into the divalent form. At the same time, the conductivity rose, in some cases as much as several decades. Measurements of the Hall effect showed that it was indeed anomalous, as if the associated carriers were positively charged. To use modern language, the divalent copper ions were contributing the equivalent of free positive charges (holes) to the conducting current. The effect of iodine vapor was reversible in the sense that the electrical conductivity dropped to a lower value when the vapor pressure of iodine was decreased, each pressure producing its own equilibrium value of conductivity.

Baedeker can be credited with lifting the standards of research in the field of semiconductors to a new high level and demonstrating clearly that, apart from surface effects of the type leading to rectification, semiconductors obey Ohm's law intrinsically. Tragically, he was called into service in the German army immediately at the outbreak of World War I in August of 1914 and died in battle the first week at the age of thirty-seven—a loss comparable in its way to that of Henry G. Mosely, another important contributor to this story, who was killed at Gallipoli in 1915 at the age of twenty-seven. Mosely had recently demonstrated, through a systematic study of X-ray emission spectra, soon after von Laue's discovery of X-ray diffraction by crystals, that Mendeleyev's atomic number describing the sequential position of a given element in the periodic chart, starting with hydrogen as number 1, is the same as the positive charge on the atomic nucleus, measured in units of the electron charge.

COLOR CENTERS IN THE ALKALI HALIDE CRYSTALS

Another sequence of important steps in the development of understanding of the behavior of electrons, both free and in traps, in insulating and semiconducting solids was made by Robert W. Pohl and his school of investigators at the University of Göttingen, starting in the 1920s. They began investigating with great care and with relative precision the electrical and optical properties of alkali halide crystals containing a stoichiometric excess of alkali metal—the so-called F centers. Such refined research, extended over many decades and augmented by that of groups in laboratories in many countries, eventually provided a firm, broad foundation for highly informative research in a number of areas of solid-state physics and chemistry, including that in the field of semiconductors.[5]

IONIC OR ELECTRONIC CONDUCTIVITY?

Activity in the field of semiconductors abated during World War I, except for special applications to coded wireless. It resumed, however, in the 1920s, developing in two important ways. First, ingenious physical chemists, such as C. Tubandt and Carl Wagner, devised methods for distinguishing between electronic and ionic conductivity in controversial cases.[6] They used techniques that made it possible to determine whether electrolytic deposition occurred at boundaries in a linear series of appropriately chosen conductors. In some cases it was found that the conductivity is actually mixed, being partly ionic and partly electronic. Quantitative analysis of electrolytic deposits made it possible to determine the ratio of the two types.

Second, C. Wagner, W. Schottky, Ya. Frenkel, and W. Jost proposed special mechanisms whereby ions could migrate in ionic crystal lattices. The processes involved depended upon the generation of vacant lattice sites or interstitial ions as a result of thermal excitation. It followed that the ionic conductivity in such crystals requires the combination of a two-step, thermally activated procedure: generation of lattice imperfections such as vacant lattice sites or interstitial ions and, coincidentally, migration made possible by the existence of such imperfections. The underlying principles involved in these developments have been tested experimentally and justified in detail in many solids. It is known, for example, that the ionic current observed in the alkali halide crystals is linked to the presence of vacancies in both the positive and negative ion lattices. In contrast, ionic conductivity in silver chloride (AgCl) depends upon the migration of interstitial silver ions.

Additions of appropriate compounds can influence the electrolytic conductivity. For example, when calcium chloride ($CaCl_2$) is added to sodium chloride, the divalent calcium ion replaces two sodium ions, creating a vacancy in the sodium ion portion of the lattice that enhances the ionic conductivity of that ion.

IMPORTANCE OF IMPERFECTIONS IN SEMICONDUCTORS

Based on experimental evidence available in 1930, B. Gudden, another of the pioneers in the systematic study of semiconductors, suggested that the electronic conductivity observed in such materials could always be traced back to some form of impurity or other defect and that they would be insulators if the imperfections were eliminated. He subsequently expounded upon this issue in a review article written in 1934.[7]

Extensive research carried out during and after World War II, particularly that on silicon and germanium, has demonstrated that Gudden's proposal, although not unreasonable for the time, is half right and half wrong. There are semiconductors in which the valence electrons of the bulk material are so tightly bound that they never contribute significantly to the electronic conductivity. In such cases, only electrons (or as will be seen, holes) that are linked to defects or impurities, and are much less tightly bound, can contribute to the conductivity. By contrast, there are many important cases in which the component of conductivity arising from the impurities or defects dominates at low temperatures but the carriers released from the bulk material dominate at elevated temperatures. This happens to be true for silicon and germanium. Each type of semiconductor must be examined separately.

INTRODUCTION OF WAVE MECHANICS

Many of the most perplexing mysteries concerning electron conductivity in metals and semiconductors were clarified by the development of wave mechanics in the second half of the 1920s by E. Schroedinger, W. Heisenberg, N. Bohr, M. Born, W. Pauli, P. Dirac, E. Fermi, and others. The new mechanics was first applied to problems involving simple atoms and molecules, such as the helium atom and the hydrogen molecule. They had evaded exact treatment in the first semiclassical theory developed by N. Bohr and extended by A. Sommerfeld just before World War I.

It was inevitable that the challenges offered by metals and other solids would soon be the object of major attention. Solutions to the outstanding mysteries hinged on two innovations. The first major advance was the discovery that electrons, like light quanta, possess wave as well as particle properties, as described by the Schroedinger equation. The second was the realization that electrons also possess the very special attribute that, in the entire assembly, only one is permitted in a given quantum state, that is, can possess a given wave function. This rule is valid for a large crystal as well as for single atoms or molecules. The exclusion principle, as it is called, was first postulated by W. Pauli as he sought to explain the way in which electrons successively occupy the electronic levels of atoms as one goes from element to element in the periodic table. At first it appeared that two electrons could be in the same state, or orbit, but the discovery of the two states of spin of the electron by S. A. Goudsmit and G. Uhlenbeck made it clear that the two electrons had to have opposite spin.

In view of the new knowledge regarding the restrictions on the way in which electrons may occupy states in a multi-electron system, H. A. Bethe and A. Sommerfeld reexamined P. Drude's free electron gas theory of metals.[8] In the old theory, it was assumed that the electrons cluster near the lowest energy state with a mean distributed energy above that state determined only by thermal agitation. In this case, as mentioned earlier, there would be a statistical spread of the order of 3RT/2 per mol of electrons, where R is the gas constant and T the absolute temperature, as in a perfect classical monatomic gas. The actual distribution of free electrons as determined by the exclusion principle, and quantified by E. Fermi and P. Dirac, changed this interpretation radically for the densities of electrons that occur in typical metals with high electrical conductivity (see diagram, page 69). Most of the electrons are densely distributed in a band several electron volts wide; only a small fraction of the electrons at the top of the band display a thermal distribution of the order of kT, where k, Boltzmann's constant, is the gas constant divided by Avogadro's number. The change explained, for example, why the electrons do not contribute as much to the specific heat of the highly conducting metals as if they were free in the classical sense. Actually, it was later demonstrated experimentally that they do contribute a small amount that increases linearly with temperature. The work of Bethe and Sommerfeld not only validated the new version of mechanics as applied to solids, but made the Drude electron gas theory of metals reputable as a semiquantitative working model of an ideal metal.

THE BAND THEORY

Maxmillian Strutt was the first individual to solve the Schroedinger equation for electrons in a periodic potential field that corresponded at least schematically to the periodic field in a crystal lattice.[9] He found that the solutions were wavelike and formed continuous bands of levels in which gaps could appear, depending upon the nature of the potential field.

Felix Bloch augmented this by developing wavelike functions constructed of series of atomic functions centered about the nuclei of the atoms in the lattice, each having a complex constant coefficient with a phase corresponding to the wavelength of a given wavelike solution.[10] Using this approximation, he was able to demonstrate that the normal temperature-dependent electrical resistivity of metals is the result of the scattering, or diffraction, of the electron waves by waves associated with the thermally excited oscillations of the atoms in the lattice. This contribution increases linearly with temperature at high temperatures when all lattice vibrational waves are excited, but falls rapidly at low temperatures where quantum effects cause the vibrational waves of higher frequency to die out. The scattering of the electron waves by imperfections or impurity atoms adds to the electrical resistivity, leading to a finite value near the absolute zero of temperature if the material is not a superconductor.

ALTERNATE VIEW OF BAND STRUCTURE

There is a sometimes convenient alternate way of looking at the energy bands found by Strutt.[11] It came into common use following the development of the method used by Wigner and Seitz (described below) to obtain more realistic three-dimensional wave functions. Imagine a lattice array of atoms that are widely separated from one another at start, so that the energy states of the electrons in the system are those of free atoms. They then exhibit typical discrete levels of the type associated with bound atomic orbits, that is, levels responsible for the sharp line emission and absorption spectra associated with free atoms. As the atoms in the extended lattice are brought closer and closer while retaining the same relative geometrical lattice arrangement, the potential fields of neighboring atoms will begin to overlap. The result is a widening of the discrete atomic levels into bands of levels that would be completely continuous for a truly infinite lattice, but which are discrete, although finely spaced, for a finite crystal because of the boundary conditions placed on the wavelike solutions to the Schroedinger equation. As the atomic levels spread into bands during lattice contraction, the electron wave functions that were originally characteristic of free atoms are modified and take on the characteristics of free electrons able to flow through the entire lattice, being modulated by the effects of the local atomic fields.

Two interesting cases from the early historical literature are presented in illustrations on pages 70 and 71. The first shows in a semiquantitative manner the development of bands from discrete atomic levels with decreasing lattice spacing in the case of crystalline sodium. It, incidentally, has a relatively close-packed lattice arrangement (body-centered cubic, that is, having atoms at the corners and centers of cubes). The second graph shows in contrast the corresponding situation for the diamond form of carbon, which has the much more open lattice structure shown on page 71. In the case of sodium, which is typical of the highly conducting metals, the overlapping of the energy level bands derived from the atomic levels is continuous at the observed lattice spacing, represented by the vertical dashed line. In contrast, the development of bands is quite different in the case of diamond. Although broadening occurs when the atomic fields of neighboring atoms begin to overlap on lattice contraction, one group of bands is depressed and becomes separated from those above it, leaving a gap in which no levels occur. In fact the lower group of bands, which have branched off, contains just enough energy levels to accommodate all the valence electrons of the carbon atoms in the crystal. The basic difference between the two cases shown in the figures can be ascribed primarily to differences in the lattice structures involved. This has been reaffirmed by later, more exact calculations.

The appearance of the gap in the representation on page 71 has great significance for the role that elements and compounds having the lattice structure shown on the same page play as semiconductors.

For the normal interatomic separation in solids, the tightly bound inner-shell electrons of a typical atom retain all but a trace of the sharply defined level structure they exhibit in the free atomic state.

METALLIC SODIUM

E. P. Wigner and F. Seitz were the first to obtain realistic wave functions for a real solid, using a method that has subsequently been generalized and improved upon for many materials.[12] In the special case of metallic sodium they found that, at the observed lattice spacing, the wave functions for the electrons in the lowest levels of the band derived from the valence electrons were almost the same as those for free electrons. The modulation caused by the potential fields near the nuclei was relatively small. This proved to be an exceptionally simple system, although an historically important one.

TYPES OF BAND STRUCTURE

It is to be expected that the electrons that occupy atomic orbitals when the atoms are widely separated will occupy extended portions of one or more bands in the quasicontinuous structure that develops as the atoms overlap (see graphs on pages 70 and 71). This is in accordance with the exclusion principle, which played such an important role in resuscitating Drude's theory based on an electron gas model of an idealized metal.

There are four, more or less fundamental cases to consider, namely those shown on page 73. In the first two (a and b), the bands of levels derived from the original atomic states as the lattice spacing decreases overlap completely, upward from the ground state associated with the valence electrons. In these cases, the upper part of the portion occupied by electrons joins continuously with vacant levels, as in the case of perfectly free electrons, for which the spectrum is entirely continuous. In the third case (c), there is a gap in the band structure, but the lower band is only partly occupied by electrons and the occupied portion of the band of levels borders empty ones, as in the first and second cases. In the fourth case (d), which has a band structure with a gap similar to that in the third, the ratio of electrons to the number of levels in the lower band is exactly one so that the lower band is completely full and the band above it completely empty.

It can be demonstrated that in the first three cases, the electrons at the upper edge of the filled region have sufficient freedom to conduct a current when an electric field is applied; hence the material is metallic. This is not so, however, in the fourth case. Complete filling of the occupied band, in compliance with the exclusion principle, deprives the electrons of the freedom that is available in the first three cases. It corresponds to an electrical insulator, at least at temperatures sufficiently low that electrons cannot be excited thermally from the filled to the empty band. Should the gap be sufficiently narrow at any given temperature that such thermally activated transitions can occur, one can expect the electrons to gain sufficient freedom to conduct a current, the conductivity rising with increasing temperatures as the number of thermally excited electrons increases. A material of this type is termed an *intrinsic* semiconductor.

The fourth case is very important for the purposes of this book since, among other things, it describes the situation for materials that can crystallize in the diamond lattice structure (see page 71), particularly, carbon, silicon, germanium, and (gray) tin. All but diamond itself exhibit pronounced intrinsic semiconductivity.[13]

The band gap for diamond, as determined by optical means, is the order of 5.7 electron volts so that intrinsic conductivity would be observed only at elevated temperatures where the diamond structure is thermodynamically unstable and reverts to graphite. The value for the gray form of tin, which has the diamond lattice structure and is stable only below 13.2° C, is very small, of the order of 0.1 electron volts, according to measurements made by G. Busch and colleagues.[14] The values for silicon and germanium are about 1.1 and 0.66 electron volts, respectively, at 300° K.

Theoretical studies, in agreement with experiment, demonstrate that the electrons that are raised from the filled band to the empty one in cases such as (d) on page 73 behave like negative charges, although they may exhibit an "effective" mass that differs from that of the free electron in a vacuum as a consequence of the detailed structure of the levels at the bottom of the empty band.

The residual electrons in the previously filled band gain freedom of motion when some of the electrons are removed from that band, thereby creating "holes" in it. Since not all the levels are filled, there is an opportunity for freedom of motion of the residual electrons near the top of the nearly filled band, which can lead to net current flow in the presence of an electric field. Thus the partly empty band can now contribute to the electrical conductivity, which may conveniently be ascribed to the presence of the holes. The current associated with their motion, however, makes an anomalous contribution to the Hall effect, as if it were the result of the motion of positive rather than negative charges. The apparent reversal of sign of the charged carriers is linked to the fact that the density of levels near the top of the filled band decreases with increasing energy as one approaches the top of the band and the electrons behave as if decelerating with increasing energy. There is a close resemblance between this behavior and that of the positron observed in radioactive decay if the latter is viewed as being produced by the removal of an electron from an infinite sea of negative energy states of the electron, as was the case in the early history of the positron. For convenience, it has become conventional to regard this anomalous form of current as being the result of the motion of holes in the nearly filled band and to treat them as virtual positive charges. Again, the effective mass of the holes may differ substantially from that of a completely free electron.

A. H. Wilson was one of the individuals most responsible for the clarification in the early 1930s of many of the properties of metals and semiconductors described here.[15] He wrote three valuable books summarizing the band theory as applied to solids. He worked on band structure calculations, as well as matters related to radar, until the end of World War II, and then became a successful business executive.

EXTRINSIC SEMICONDUCTORS

Most of the applications of semiconductors that will be described in later chapters of this book center about forms of the materials, particularly silicon and germanium, to which foreign atoms have been added in controlled amounts. The following brief account of some of the effects induced by such additions will benefit those not already intimately familiar with the subject.

We have seen that the electronic conductivity and other properties of semiconductors are very sensitive to impurities and other forms of imperfection. A part of this sensitivity can readily be accounted for in the framework of the band theory of solids using elemental semiconductors such as silicon and germanium as illustrative cases. Suppose, for example, that a small fraction of the silicon atoms are more or less randomly replaced by elements from the fifth column of the periodic table, such as phosphorus or arsenic. Four of the five valence electrons of the additive atoms can be expected to be close stand-ins for those of the silicon atoms they replace, although there will be some degree of misfit, and that occupy levels in the filled band. Since that band can accommodate only four of the five electrons associated with the additional atom, in keeping with the rules of quantum statistics, the fifth electron must accept a level (or state) that represents a compromise between characteristics of the levels of the upper empty band and the fact that it is linked to one of the substituted atoms, which has an attractive electrostatic potential field. As a result, it occupies a discrete level in the interband gap that lies just under the empty band. Any electron occupying such a level will be distributed over a volume, but it will be centered about the substitutional atom to which it is attracted. A typical example of such a level is shown in (b) on page 74. Being close to the empty band, the electrons that may occupy such levels at very low temperatures can be expected to be excited into the empty band at a much lower temperature than the electrons in the filled band. Such high-lying discrete energy levels are commonly called *donor levels*.

It follows that any impurities, whether substitutional or interstitial, which contribute occupied discrete levels of this kind will also contribute to the electrical conductivity with a normal Hall effect at a lower temperature than the electrons in the filled band, although the latter can dominate at sufficiently high temperatures because of their greater number. The conductivity associated with such additive atoms is termed *extrinsic*.

If instead of substituting a quintivalent atom, we were to substitute a trivalent one such as boron or aluminum, the three valence electrons could enter the filled band as stand-ins. In this case, however, the substitutional atoms do not provide a sufficiently large number of electrons to occupy all of the possible levels in the occupied band permitted by quantum statistics. As a result the modified system develops a sequence of unoccupied discrete levels that are just above the occupied valence band. Any electron that happens to occupy such a level will be bound and localized about one of the substituted atoms. It will partake of some of the characteristic properties of that atom and of the levels in the occupied band since the discrete levels are closely associated with it. It follows that such trivalent additions can be expected to introduce energy levels of the type shown in (c) on page 74, which lie near the filled band and are unoccupied at very low temperatures. Or, in terms of a commonly used convention, they can be viewed as if occupied by a hole. At some level of thermal excitation, a fraction of the vacant discrete levels will become occupied by electrons from the filled band, thereby generating holes in the latter that can transport an electric current displaying the anomalous Hall effect. Again this type of extrinsic conductivity can be expected to appear at much lower temperatures than the intrinsic conductivity. The associated unoccupied energy levels are called *acceptor levels*.

As we shall see in chapter 13, which focuses on research at the Bell Telephone Laboratories, extrinsic conductivity of the type shown in case (b) on page 74 apparently first came to be called n-type at the laboratories because the effective carriers behave as if negatively charged, whereas those that conduct by means of holes (case c) were designated p-type. In any event the terminology gained immediate use throughout the community associated with semiconductors.

Since natural crystals may contain large admixtures of impurities of atoms of different kinds that are not uniformly distributed though the lattice, it is not surprising that, as semiconductors, they exhibit great variability. In fact, some regions could be predominantly n-type and others p-type. Any hope for gaining control of the properties of a given semiconducting material will require, among other things, that great attention be given to chemical composition. As we shall see, the most advanced modern technology based on the use of semiconductors requires chemical analysis and control to the level of parts per billion (10^{-9}) or better.

The large-area cuprous oxide semiconductors that were used for rectification of alternating currents, starting in the 1920s, were made by processing discs of copper sheet made from copper prepared at commercial smelters. Since the composition of the copper varied from one source to another and from time to time, the manufacturing process required continuous, vigilant testing to ensure appropriate quality.

THREE-FIVE (III-V) COMPOUNDS

The characteristic semiconducting properties of silicon and germanium described in the foregoing pages were revealed for the most part as a result of research carried out in connection with radar during World War II, as will be discussed in chapter 11. In order to extend the family of possibly useful materials, which exhibit both intrinsic and extrinsic semiconductivity, a great deal of attention was given in the postwar period to compounds such as gallium arsenide (GaAs) formed of a combination of elements from the third and fifth columns of the periodic chart, which have the same basic lattice structure as diamond, silicon, germanium, and gray tin and which possess a comparably small energy gap. H. Welker, in particular, made a special study of a range of such compounds, some of which have found auxiliary roles in electronics.[16] At first it appeared that some of these compounds might seriously challenge the roles of silicon and germanium in the development of transistors (see chapter 14), However, the severe demands placed on materials that can be used in integrated circuits, as well as advances made in silicon process technology, have made this possibility seem unlikely, although they have found a valuable place in devices such as light-emitting diodes (see chapter 5), infrared detectors, and in specialized high-speed integrated circuits.

(Above left) Georg Busch of the Swiss Federal Institute in Zurich (ETHZ), one of the pioneers in semiconductor research. (Courtesy of Busch.)

(Above right) Humphrey Davy, the first director of the Royal Institution. He recognized Faraday's remarkable gifts and gave him free rein. (Courtesy of the American Institute of Physics Emilio Segrè Visual Archives.)

(Right) Michael Faraday as a youthful, but already well-recognized, scientist. (Engraving by J. Cochran of a painting by H. W. Pickersgill; courtesy of the American Institute of Physics Emilio Segrè Visual Archives.)

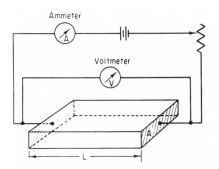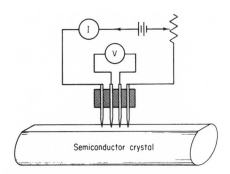

Schematic view of the four-probe method for determining the electrical conductivity of a semiconductor (right-hand figure). The left-hand diagram shows the "direct" method of measuring the conductivity. That method can give false results if the contacts are rectifying or otherwise contribute to the resistance. (Courtesy of Robert G. Hibberd. Reprinted by permission of Texas Instruments Incorporated.)

E. H. Hall, who, with the encouragement and support of Henry Rowland, discovered that a current-carrying conductor develops a transverse electromotive force in the presence of a magnetic field perpendicular to the flow of current. (Courtesy of the American Institute of Physics Emilio Segrè Visual Archives. Photo by Bachrach.)

Paul Drude, who developed the free electron theory of metals at the turn of the twentieth century using classical Maxwell-Boltzmann statistics. He encountered serious contradictions with experimental observations. (Courtesy of the Deutsches Museum, Munich.)

Johan Königsberger, another of the early scientists to speculate upon the nature of semiconductors. He concluded correctly that the carriers were frozen-in at sufficiently low temperatures and require an activation energy to become mobile. (Courtesy of Georg Busch, who received a copy of the photograph from Königsberger's daughter, Mrs. M. J. Loveday-Königsberger.)

Karl Baedeker, who made the first systematic scientific study of the chemical and physical properties of semiconductors, starting in 1909. He used chemically pure metal foils and converted them to semiconducting compounds, such as cuprous iodide. (Courtesy of the Archives of the University of Jena.)

Robert W. Pohl, who, starting in the 1920s, devoted a major portion of the research in his physics institute at Göttingen to the study of the electrical and optical properties of highly purified alkali halide crystals containing a stoichiometric excess of alkali metal atoms (F centers), as well as other additives. (Courtesy of the Walter Brattain Collection of the Emilio Segrè Visual Archives of the American Institute of Physics.)

Ya. Frenkel, the brilliant Soviet scientist who made contributions to many fields of physics and theoretical chemistry. He developed one of the first theories concerning the nature of the thermally induced lattice defects responsible for ionic migration in ionic crystals. (Courtesy of the American Institute of Physics Emilio Segrè Visual Archives, Frenkel Collection.)

Hans A. Bethe, who, with Arnold Sommerfeld, resuscitated the Drude free-electron theory of metals with the use of quantum statistics. (Courtesy of the American Institute of Physics Meggers Gallery of Nobel Laureates.)

Arnold Sommerfeld *(left)* with Niels Bohr. Both were major figures in the development of quantum mechanics. (Courtesy of the American Institute of Physics Emilio Segrè Visual Archives, Margrethe Bohr Collection.)

The Fermi-Dirac distribution of electrons in the Drude free-electron theory of metals as developed by Bethe and Sommerfeld. The shaded area of (a), which extends for several electron volts, 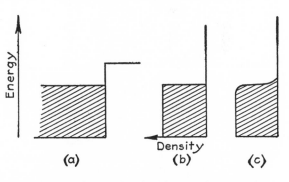 represents the range of levels in the conduction band of the metal occupied in accordance with the Pauli exclusion principle. The confining potential barrier at the surface of the metal is shown. The distribution function (b) is for the absolute zero of temperature; the right-hand one (c) is for a finite temperature, the range of thermal excitation at the top of the occupied range being kT, where k is Boltzmann's constant and T is the absolute temperature.

Felix Bloch, who developed an approximate family of traveling waves in a metal from atomic wave functions. He used the system to demonstrate that the temperature-dependent electrical resistivity of normal metals is the result of the diffraction of the electron waves by the lattice vibrational waves. (Courtesy of the American Institute of Physics Meggers Gallery of Nobel Laureates.)

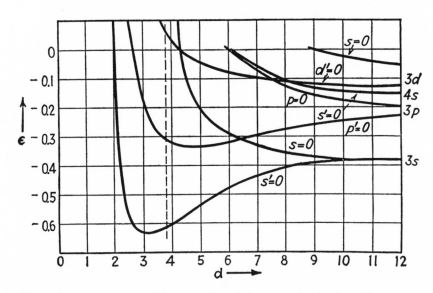

Schematic representation of the spreading of the atomic levels of atomic sodium, organized in a lattice array, as the interatomic spacing is decreased to distances at which valence electrons on neighboring atoms overlap. (After J. C. Slater, *Physical Review* 45 [1934]: 794.)

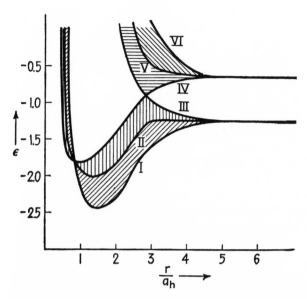

An early, semiquantitative calculation, analogous to that shown on page 70, but for the case of diamond, which has the relatively open lattice structure shown below. In this case, a distinct gap appears in the structure of the broadening bands associated with the valence electrons as the atomic fields of neighboring atoms begin to overlap. In a historical sense, this diagram implicitly contains one of the first intimations that crystals of elements such as silicon and germanium, as well as compounds having the diamond lattice structure, might be intrinsic semiconductors. (After G. E. Kimball, *Journal of Chemical Physics* 3 [1935]: 560.)

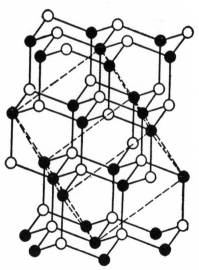

The diamond lattice structure, observed in diamond, silicon, germanium, and gray tin. In these cases, the white and black circles correspond to identical atoms. The lattice is also found in compounds such as gallium arsenide (GaAs) and zinc sulfide (in the sphalerite form). In such cases the two types of circles correspond to the two different atomic species.

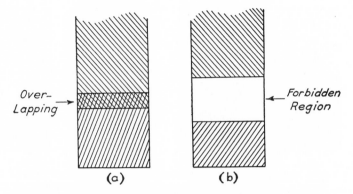

Two basic patterns of bands that can develop in a solid. In the left-hand case, the bands overlap from the lowest valence electron level upward; in the right-hand case, a gap appears.

Eugene Wigner, the very versatile Hungarian-born scientist-engineer who worked in many fields. Although he was trained as a chemical engineer and was principally responsible for the design of the first large plutonium-producing reactors, he made many fundamental contributions to quantum theory, as well as to nuclear chemistry and physics. He is shown here with his granddaughters, Margaret *(front)* and Mary Upton. (Courtesy of Martha Upton.)

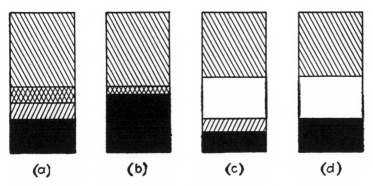

Four possible ways in which the energy bands may be occupied in principle by the valence electrons. In cases (a) and (b), the bands overlap continuously; the solids are metals. Although there is a gap in case (c), the band containing the valence electrons is only partly filled, so it also corresponds to metallic behavior. In case (d), the lower, separated band is completely full, the one above it completely empty; this represents an electrical insulator, or an intrinsic semiconductor if the band gap is sufficiently small to permit thermal excitation from the filled to the empty band.

A. H. Wilson, a highly productive pioneer in the early development of the band theory of solids. (Courtesy of the Meitner-Graf Studio and Robert Cahn.)

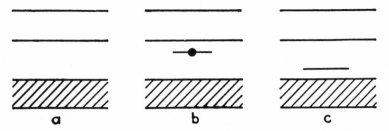

Case (a) represents the ideal intrinsic semiconductor. Case (b) corresponds to a case in which a silicon atom has been replaced by an element such as phosphorus from the fifth column of the periodic table, bringing an extra electron that occupies a discrete, localized level in the forbidden gap, near and just below the bottom of the empty band. Case (c) corresponds to that of a specimen of silicon in which an element from the third column, for example, aluminum, has been substituted. A discrete unoccupied level is created in the forbidden gap, near and just above the top of the filled band. This corresponds to the case of an extrinsic p-type semiconductor. In a commonly used convention, the unoccupied discrete level in case (c) is viewed as being occupied by a hole that can make a transition to the filled band and become mobile.

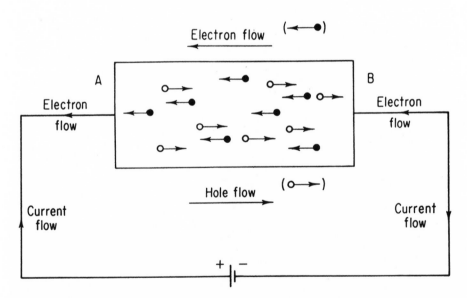

Schematic diagram showing the opposite direction of motion of electrons and holes when an electric field that drives the electrons to the left is applied. (Courtesy of Robert G. Hibberd. Reprinted by permission of Texas Instruments Incorporated.)

NOTES

1. Among the many good books on aspects of solid-state physics are the following: Leon Brillouin, *Quantenstatistik* (Berlin: Springer, 1931); A. H. Wilson, *Semiconductors and Metals* (Cambridge: Cambridge University Press, 1939); idem, *The Theory of Metals*, 2d ed. (Cambridge: Cambridge University Press, 1954); N. F. Mott and R. W. Gurney, *Electronic Processes in Ionic Crystals* (Oxford: Oxford University Press, 1940); F. Seitz, *The Modern Theory of Solids* (New York: McGraw-Hill, 1940); W. B. Shockley, *Electrons and Holes in Semiconductors* (New York: Van Nostrand, 1950); H. K. Henisch, ed., *Semiconducting Materials* (New York: Academic Press, 1951); W. B. Shockley et al., eds., *Imperfection in Nearly Perfect Crystals* (New York: Wiley, 1952); A. H. Cottrell, *Dislocations and Plastic Flow in Crystals* (London: Oxford University Press, 1953); R. Peierls, *The Quantum Theory of Solids* (Oxford: Oxford University Press, 1955); H. G. Van Bueren, *Imperfections in Crystals* (Amsterdam: North Holland Publishing Co., 1960); J. H. Schulman and W. D. Compton, *Color Centers in Solids* (New York: Pergamon Press, 1962); D. Pines, *Elementary Excitations in Solids* (New York: W. A. Benjamin, 1963); N. A. Goryunova, *The Chemistry of Diamond-Like Semiconductors* (Cambridge, Mass.: MIT Press, 1965); C. T. Tomizuka and R. M. Emrich, *Physics of Solids at High Pressures* (New York: Academic Press, 1965); A. S. Grove, *Physics and Technology of Semiconducting Devices* (New York: Wiley, 1967); W. B. Fowler, ed., *Physics of Color Centers* (New York: Academic Press, 1968); R. A. Levy, *Principles of Solid State Physics* (New York: Academic Press, 1968); R. G. Hibberd, *Solid-State Electronics*, Texas Instruments Electronics Series (New York: McGraw-Hill, 1968); M. Doyama and S. Yoshida, *Point Defects* (Tokyo: University of Tokyo Press, 1977); L. Hoddeson et al., eds., *Out of the Crystal Maze* (New York: Oxford University Press, 1992); C. Kittel, *Introduction to Solid State Physics* (New York: Wiley, 1996). Those desiring a very broad view of the field can find it in the series of volumes titled *Solid State Physics*, ed. H. Ehrenreich and Frans Spaepen (New York: Academic Press).

2. See Georg Busch, "Early History of the Physics and Chemistry of Semiconductors," *Condensed Matter News* 2, no. 1 (1993): 15.

3. A biography of E. H. Hall appears in the *Biographical Memoirs of the National Academy of Sciences*, vol. 21 (1941), p. 73.

4. For discussions by K. Baedeker, see *Physik. Zeitschrift* 9 (1909): 431; *Annalen d. Physik* 29 (1909): 566; *Physik. Zeitschrift* 13 (1912): 1080; *Die Elektrischen Erscheinungen in Metallischen Leitern* (Braunschweig: Vieweg, 1911).

5. Reviews of such work may be found in the sources listed in n. 1.

6. Ibid.

7. See B. Gudden, *Ergebnisse der Exakten Naturwissenschaften* 13 (1934): 223.

8. A. Sommerfeld, *Zeitschrift f. Physik* 47 (1928): 1; A. Sommerfeld and H. Bethe, "Elektronentheorie der Metalle," *Handbuch der Physik*, vol. 24, pt. 2 (Berlin: Springer, 1934).

9. Reviews of such work may be found in the sources listed in n. 1.

10. Ibid.

11. See J. C. Slater, *Physical Review* 45 (1934): 794.

12. Reviews of such work may be found in the sources listed in n. 1.

13. See Goryunova, *Chemistry of Diamond-Like Semiconductors*, n. 1, above; see also the discussion on silicon in chap. 2, n. 14, in the present book.

14. G. Busch, J. Wieland, and H. Zoller, "Electronic Properties of Grey Tin," in *Semiconducting Materials*, ed. H. K. Henisch (London: Butterworth Scientific Publications, 1951), p. 188, n. 1.

15. See Wilson, *Semiconductors and Metals* and *The Theory of Metals*.

16. H. Welker and H. Weiss, *Solid State Physics*, vol. 3 (New York: Academic Press, 1957), p. 1. An account of some of Welker's research on diode rectifiers appears in *Jahrbuch der deutschen Luftfahrtforschung* 3 (1941): 63.

FIVE

RECTIFICATION

$\mathscr{I}$T IS NOW POSSIBLE to discuss the type of rectification first observed by Ferdinand Braun and subsequently used in coded wireless.[1] The top figure on page 82 shows the energy levels, occupied and empty, for a typical isolated metal. The quantity W, the *work function,* is the energy required to remove an electron from the top of the occupied region into a vacuum, whereas W' is the counterpart for the bottom of the band. In actual practice these values will depend upon the state of the surface since, as will be seen in chapter 14, the surface may possess foreign impurity atoms that also generate a contribution to the dipole layer that affects W.

Diagram A on page 82 shows an equivalent representation of an isolated extrinsic semiconductor in which the donor levels are just below the conduction band. Diagram B, on the same page, represents the levels in a typical metal, such as that illustrated on page 69. In this case, it is assumed that the work function W_s for the donor levels in the semiconductor is less than that for the metal, W_m. When the two are brought together and to equilibrium (see page 82), some of the electrons in the trapped donor levels can reduce the energy of the system by flowing into the metal, thereby creating a negatively charged layer on the metal side of the boundary, and a positive, so-called *depletion layer,* of much greater width on the semiconductor side. In the case shown, in which W_m is larger than W_s, the electrons in the metal encounter the steep barrier $W_m - W_s$ opposing migration if the metal serves as the cathode. An electron that does flow into the semiconductor from the metal must be highly excited, so as to have sufficient energy to go over the barrier, or must tunnel through in the manner permitted by wave mechanics. If the situation is reversed and the semiconductor becomes the cathode, the bottom of the conduction band is raised and the electrons in the semiconductor, now flowing toward the metal, encounter a much lower barrier since the slope of the lower part of the conduction band in which the depletion layer occurred is flattened. In an extreme case (see page 83) the band is completely flattened and the flow is essentially unimpeded by the barrier. It follows that the current-voltage curve for the system is asymmetric and that the metal-semiconductor junction is rectifying. It constitutes the "blocking layer" mentioned in chapter 1.

In this example, which is typical of the situation for a metal-semiconductor diode rectifier, the rectification is related to two basic properties of the components: The work function for the metal is larger than that for the donor levels in the semiconductor; the potential drop in the depletion layer of the semiconductor that occurs in the equilibrium state (page 82) can be altered by reversing the applied field because that layer is much wider in the direction perpendicular to the junction than the oppositely charged layer in the metal.

This interpretation of the metal-semiconductor rectifying junction was developed independently by Nevill F. Mott and Walter Schottky in the 1930s. Exact quantification of their theories proved difficult, however, because electric dipole layers at the metal-semiconductor interface arising from extraneous sources can complicate the situation.

The upper diagram on page 84 shows schematically the corresponding equilibrium distribution of bands and levels for the junction of a metal and an extrinsic p-type semiconductor. The plus signs (+) and minus signs (−) represent the dipole space-charge distribution at the boundary, narrow in the metal and relatively broad in the semiconductor.

One has here the choice of describing the rectifying properties in terms of the behavior of the electrons or holes. Let us focus on the electrons first. If the potential of the semiconductor is made more positive relative to the metal, in the conventional sense, its bands and levels will be lowered relative to those of the metal, and become flatter in the vicinity of the junction. This will permit some electrons on the metal side of the boundary to enter the not quite full band of the semiconductor, the voids in the latter being a result of the thermal excitation of electrons from that band to the empty nearby acceptor levels. Or, viewed from the alternative standpoint of the hole schema, holes from the nearly filled band of the semiconductor will be able to migrate into the occupied levels of the metal. The more positive the potential difference between the semiconductor and metal, the larger is the current flow.

If the relative potential between the two sides is now reversed—the potential of the semiconductor becoming more negative, so that the bands and levels in the semiconductor are raised further relative to those in the metal—the slope of the nearly filled band of the semiconductor near the metal junction will become steeper and the transfer of charge across the metal-semiconductor boundary will become more difficult. In principle, electrons in the semiconductor that may have gained sufficient thermal energy to make the transition from its nearly filled to its empty band might, under the proper circumstance, make the further transition to the metal. This, however, would require a substantial activation energy, akin to that needed for the transition from the metal to the semiconductor in the case shown on page 82. In summary, the system is rectifying, although the direction of rectification has the opposite sign from that observed for the junction between the metal and an n-type extrinsic semiconductor.

The basic principles described here apply to the case of the point-contact diode using a metal whisker—a system complicated only by the fact that the current flowing from the tip of the whisker spreads out into the semiconductor after leaving the metal, or contracts at the junction when flowing in the reverse direction.

THE P-N JUNCTION

Another very important form of rectifying unit is obtained by creating a continuous junction of n-type and p-type forms of a given semiconductor, such as either silicon or germanium. The idealized situation for the separated semiconductors is shown at the bottom of page 84. When the two are created as a continuous unit and are in electrical equilibrium, some of the electrons in the n-type specimen will migrate to fill vacant levels in the p-type specimen, causing a dearth of electrons in a layer near the surface of the first and an excess of electrons (or dearth of holes) in a corresponding layer in the second. An electrostatic dipole, or space-charge, layer will result. The relative widths of the bands possessing a dearth of electrons in the n-type specimen and an excess in the p-type will depend upon the relative densities of additive atoms, that is, *dopants,* in the two specimens. The illustration on page 85 shows schematically the new disposition of charges and the associated potential for the previous diagram, an example in which the densities of electrons and holes are about the same in the two specimens—a situation that need not be true in actual cases.

The lower diagram on page 85 displays an alternate view of the charge distribution in an idealized p-n junction, whereas the diagram on page 86 illustrates schematically the manner in which the current is divided between the migration of electrons and holes on either side of the junction for the case of the applied potential illustrated in the diagram.

If a negative voltage that raises the height of the n-type region in the upper diagram on page 85 is applied, electrons will be able to flow more readily to the right in that region and holes to the left in the p-type region. If the potential is reversed, the barrier to the net flow of current is raised. It follows that the junction is rectifying. If, in fact, the reversed voltage is sufficiently high, electrons will be driven to the left and holes to the right, depleting the junction area of carriers except to the extent that electron-hole pairs are generated thermally or by some form of high-field breakdown.

ELECTRON-HOLE RECOMBINATION

Should one make the potential of the n-type semiconductor in the p-n junction sufficiently negative relative to the p-type, the electrons and holes will be driven toward one another and into the dipole layer where they will recombine, producing either lattice vibrations, corresponding to heat, or photons. Which of the two occurs depends upon the material and other factors,[2] such as the relative period of time required for the electron-hole pair to recombine with the emission of light, if one depends upon spontaneous emission of incoherent light. Unlike free atoms, such recombination with the emission of a light quantum may be complicated in solids depending upon the structure of the two bands containing the electron and hole. In some cases (termed indirect transitions) a light quantum can be emitted only with the special cooperation of lattice vibrations, thereby lengthening the time for the emission of a light quantum. In other cases, as is true for excited F-centers in the alkali halide crystals studied by Pohl (see chapter 4), no such special cooperation is needed. Such transitions are termed direct. It appears that the recombination of electrons and holes is indirect and hence relatively slow in both silicon and germanium. This fortunately favors the lifetime of minority carriers in transistors (see chapter 14). The opposite appears to be the case in many of the three-five compounds, such as gallium arsenide. Recombination with the emission of light is usually direct in these compounds, and hence more highly favorable.

The first practical, visible, light-emitting diode was developed by N. Holonyak in 1962.[3] It emits red light and is based on a modified gallium arsenide semiconductor, actually an "alloy" of gallium arsenide and gallium phosphide, in which a fraction of the arsenic has been replaced by phosphorus. Both the range of wavelengths available with such diodes made of alloys of elements from the third and fifth columns of the periodic chart and their emission efficiency have been extended as a result of continuous research over the past three decades. For example, an indium gallium nitride compound that emits in the blue is now available for further development. An aluminum gallium indium phosphide alloy that emits in the red-orange range of the spectrum possesses an emission efficiency of the order of forty lumens per watt, comparable to that of a one-kilowatt mercury lamp. The success of these developments, which form the basis for the present-day field of electro-optics, makes it clear that the extended research on three-five compounds over the years has not been in vain.

It should be noted that electrons and holes can be injected into the light-emitting semiconductor from other semiconductors, so-called confining layers, which are in intimate electrical contact with the emitter. Significant advantage can be gained from such heterosystems if there is good optical matching and if the supplier or suppliers of electrons and holes have larger band gaps than the light-emitting compound. It may be easier for the light to escape from the device through the more transparent partners than from the primary light source without such optical matching. The development of an understanding of carrier injection, described in chapter 14, has been of substantial value for the advancement of this field.

If two opposing reflectors are introduced into a volume of the region where electron-hole recombination readily takes place with the production of light, the system can be converted into a laser, which depends upon coherent induced emission. This is the basis for the operation of the semiconducting diode laser. The diagram on page 87 shows the configuration of components in a recently developed gallium nitride laser.[4]

THE PHOTOVOLTAIC EFFECT

If a p-n junction is irradiated with light quanta having a range of energies that can raise electrons from the filled to the empty band, that is, produce electron-hole pairs, the charges will be driven in opposite directions by the dipole field, generating a photon-induced current in a connecting lead when the two halves are short-circuited. In an open circuit the electron-hole pairs will partly neutralize the potential across the dipole layer to a degree that depends upon the level of illumination. This will alter the potential between the two halves of the unit—a phenomenon called the photovoltaic effect, which is the basis for the use of semiconductors in photographic exposure meters and some solar cells.

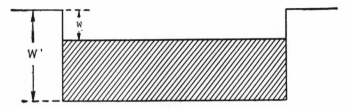

The occupied levels in a typical metal. The surface barrier, which involves a retaining dipole layer, appears on both sides. W is the work (energy) required to remove the uppermost electrons in the occupied portion of the levels (crosshatched area) to the exterior of the metal, the so-called *work function*. W' is the depth of the occupied region of levels as measured from the outside.

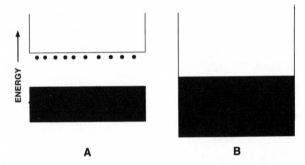

Diagram A represents schematically the levels in a typical extrinsic n-type semiconductor, as described in the previous chapter, whereas diagram B represents the levels for a typical metal, analogous to that illustrated on page 69.

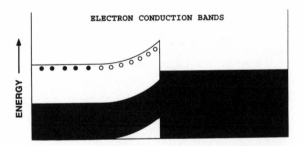

The junction barrier that develops when the metal and semiconductor shown above are placed in intimate contact and are at equilibrium. In this case, it is assumed that some electrons have been transferred from the semiconductor to the metal, creating an electron-deficit volume (also known as a depletion or inversion region) in the former and a much narrower compensating charge layer, not shown, near the surface of the metal.

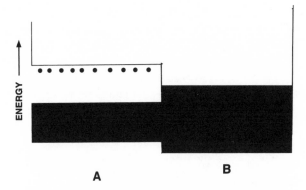

A case in which the potential of the n-type semiconductor has been raised sufficiently (that is, made sufficiently negative) that electrons can flow easily from the semiconductor to the metal. This represents the high-conductivity, or "forward bias," regime of the rectifying junction.

Nevill Mott, who proposed the junction model of metal-semiconductor rectification at the same time as Walter Schottky. (Courtesy of the American Institute of Physics Niels Bohr Library and Sir Nevill Mott. Lotte Meitner-Graf photograph.)

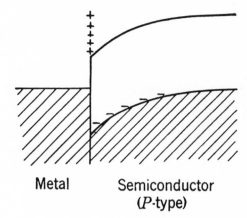

The equilibrium distribution of levels and charge in the contact between a metal and an extrinsic p-type semiconductor. Some electrons have lowered their energy by making a transition from the top of the occupied band of the metal to the vacant acceptor levels of the semiconductor, thereby creating a dipole layer. (Courtesy of H. C. Torrey and C. A. Whitmer, from *Crystal Rectifiers*.)

Metal   Semiconductor
(*P*-type)

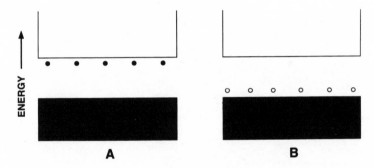

Schematic diagram showing the relative positions of occupied and unoccupied levels for an n- and a p-type extrinsic semiconductor when the two are not joined.

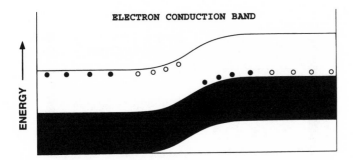

ELECTRON CONDUCTION BAND

ENERGY

The equilibrium state when the two extrinsic semiconductors shown at the bottom of page 84 are chemically and physically joined, perhaps by interdiffusion, to produce a p-n junction. Electrons have been transferred from the n-type to the p-type specimen, generating an extended dipole layer and causing a relative shift in energy levels between the two halves.

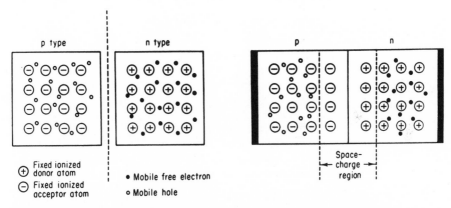

p type   n type

⊕ Fixed ionized donor atom
⊖ Fixed ionized acceptor atom

• Mobile free electron
○ Mobile hole

p   n

Space-charge region

Alternate cross-sectional representations of the situations shown in the previous two diagrams. (Note: the left-right positions of the two halves are the reverse of those shown in those diagrams.) The left-hand diagram displays the charge distribution in the two halves prior to contact; the right-hand one the formation of a dipole layer after contact. (Courtesy of Robert G. Hibberd. Reprinted by permission of Texas Instruments Incorporated.)

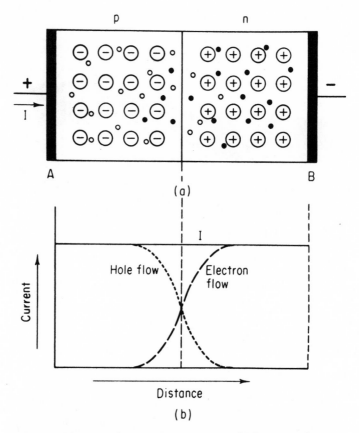

Schematic diagram showing the manner in which current flow in the vicinity of a junction is divided between electrons and holes when the potential difference between the two halves is as shown. (Courtesy of Robert G. Hibberd. Reprinted by permission of Texas Instruments Incorporated.)

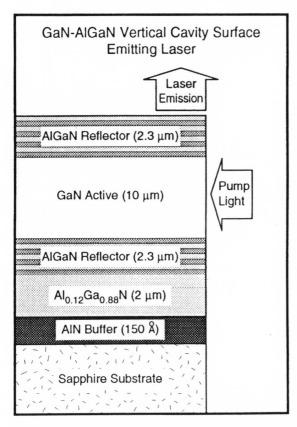

GaN-AlGaN Vertical Cavity Surface Emitting Laser

Laser Emission

AlGaN Reflector (2.3 μm)

GaN Active (10 μm)

Pump Light

AlGaN Reflector (2.3 μm)

Al$_{0.12}$Ga$_{0.88}$N (2 μm)

AlN Buffer (150 Å)

Sapphire Substrate

Schematic arrangement for producing a light-emitting diode laser, in this case, one based on the light-emitting properties of gallium nitride. The nitride is excited by radiation in the near ultraviolet, such as 3370 Angstrom units, and emits between about 3350 and 3800 Angstrom units. The light-emitting semiconductor is sandwiched between two composite, reflecting mirrors composed of an aluminum-gallium-nitride compound. The system is mounted upon a sapphire base. The numbers shown are the physical dimensions in microns (μm) of the components. (Courtesy of J. M. Redwing, D. A. S. Loeber, N. G. Anderson, M. A. Tischler, and J. S. Flynn and the American Institute of Physics.)

NOTES

1. For background, see chap. 4, n. 1.

2. We are indebted to both Nick Holonyak and Turner Hasty for very informative discussions of this topic.

3. See N. Holonyak Jr., D. C. Jillson, and S. F. Bevacqua, "Halogen Vapor Transport and Growth of Epitaxial Layers of Intermetallic Compounds and Compound Mixtures," presented at a conference of the American Institute of Metallurgical Engineers, Los Angeles, Aug. 1961; this was published subsequently in *Metallurgy of Semiconductor Materials,* vol. 15, ed. J. B. Schroeder (New York: Interscience Publishers, 1962), p. 49; see also N. Holonyak Jr. and S. F. Bevacqua, "Coherent (Visible) Light Emission from Ga(As$_{1-x}$P$_x$) Junctions," *Applied Physics Letters* 1 (1962): 82.

4. For discussion of this development, see, for example, J. M. Redwing, D. A. S. Loeber, N. G. Anderson, M. A. Tischler, and J. S. Flynn, *Applied Physics Letters* 69, no. 1 (1996): 1. A recent summary of several families of light-emitting semiconductors is given in the review paper by A. V. Nurmikko and R. L. Gunshor, "Physics and Device Science in II-IV Semiconductor Visible Light Emitters," in *Solid State Physics,* vol. 49, ed. H. Ehrenreich and F. Spaepen (New York: Academic Press, 1995), p. 205.

SIX

RADAR

$\mathcal{T}$HE SILICON STORY comes to the fore in a major way with the development of microwave technology and radar in the 1930s.[1] The topic of radar has a long prehistory going back to the early days of wireless, but many factors intervened to delay the ultimate development. As early as 1904, Christian Hülsmeyer developed a primitive but effective radar for shipboard use employing electromagnetic wave generators of the period.[2] His system used meter-long Hertzian waves with separate transmitters and receivers that were physically mounted on the same support. He was stimulated by the desire to prevent the large number of serious ship collisions occurring in the North Sea and elsewhere in periods of heavy fog. In Rotterdam harbor, he obtained clear reflections from metallic ships three kilometers away. In spite of his remarkable success, with all of the interest it provoked and the promises for future evolution, neither the German navy nor any commercial organization expressed a desire to finance further development. The topic remained dormant for the next twenty years. Hülsmeyer's work and interests were, unfortunately, far in advance of general comprehension of their importance.

French engineers developed a modern version of Hülsmeyer's system in the 1930s to protect the gigantic passenger liner *Normandie* from collisions with icebergs. They found that it did not work as well in heavy seas as in relatively quiet waters because of the reflections from sea waves, so-called sea clutter.

The issue of radar surfaced again in 1922 when the basic technology had advanced much further. Without citing Hülsmeyer's work, Marconi, in an address to the Institution of Radio Engineers, pointed out the potential of echo location with electromagnetic waves.[3] By this time, the development of vacuum tube technology was in full swing and many individuals were experimenting with radio waves above the megahertz range. It was commonplace for such observers to detect reflected waves from large metallic objects. Beat frequencies associated with Doppler shifts in the reflected wave from a moving object such as a large motor vehicle were also commonly observed.

One major stimulus for the ultimate development of radar emerged from scientific and technical interest in the detailed characteristics of the ionized layers of the upper atmosphere generated by solar radiation. Such interest was a truly international one in which many individuals in the technically advanced countries participated in the 1920s and 1930s. Much of the work was carried out in the 10-meter range of wavelengths (30 megahertz). For example, G. Breit and M. Tuve of the Carnegie Institution in Washington, D.C., working with pulsed 70-meter waves, observed the reflections on an electron oscilloscope, a very sophisticated procedure for its time.[4] Their pulse length was of the order of a millisecond; their baseline from emitter to receiver over which the reflections were observed was 12 kilometers. Most others used continuous wave techniques, depending upon reflected beams and triangulation to determine the heights of the reflecting layers and the frequency-dependence of reflectivity.

It was soon observed that one could detect echoes from overhead aircraft, particularly when metal-bodied planes became common. This stimulated military interest in what ultimately came to be called radar detection.[5] Such interest proved to be particularly strong in the larger European countries, such as Britain, France, Germany, and the Soviet Union, although, as will be seen in chapter 11, the United States Navy eventually became productively involved in pursuing the technology in the mid-1930s.

Christian Hülsmeyer, who developed a continuous-wave radar for shipboard use in 1904. He is shown here with some of his equipment. He was interested in preventing collisions between ships in fog and in crowded conditions, such as in harbors. The system was reasonably effective and, by process of evolution, could have made modern radar feasible much sooner. It did not, however, attract adequate financial backing. (Courtesy of the Deutsches Museum, Munich.)

Merle Tuve, who with Gregory Breit developed a pulsed echo system in 1924 for studying the layers of the ionosphere. They were assisted by colleagues at the Naval Research Laboratory. The pulse length was about a millisecond and marginal for the observations. Almost a decade was required to develop more effective pulsed equipment. (Courtesy of the American Institute of Physics Emilio Segrè Visual Archives.)

Gregory Breit *(right)*, a theoretical physicist who contributed to many areas of science. He and Merle Tuve carried out the first study of the ionosphere with pulsed beams in 1924, a landmark investigation. The other two individuals shown here are Charles G. Darwin *(left)*, the grandson of the famed biologist, and Llewellyn H. Thomas *(center)*. (Courtesy of the American Institute of Physics Emilio Segrè Visual Archives, Goudsmit Collection.)

NOTES

1. For general and detailed references on radar, see Ulrich Kern, "Die Enstehung des Radar Verfahrens: Zur Geschichte der Radar Technik bis 1945" (Thesis, University of Stuttgart, 1984); Henry E. Guerlac, *Radar in World War II*, 2 vols. (New York: American Institute of Physics, 1987); and idem, "The Radio Background of Radar," *Journal of the Franklin Institute* 250 (1950): 285. See also the 28-volume collection *The Radiation Laboratory Series,* ed. Louis N. Ridenour (New York: Massachusetts Institute of Technology and McGraw-Hill, 1948). For an excellent popularized version of the general history of radar, see Robert Buderi, *The Invention That Changed the World* (New York: Simon and Schuster, 1996).

2. See particularly the account of Hülsmeyer's work in Kern, "Die Enstehung des Radar Verfahrens." Mention is also made in Guerlac's "Radio Background of Radar."

3. Marconi's speech appeared in print as "Radio Telegraphy," *Proceedings of the Institution of Radio Engineers* 10 (1922): 215.

4. See G. Breit and M. Tuve, *Physical Review* 28 (1926): 554.

5. See E. B. Callick, *Metres to Microwaves* (London: Institution of Electrical Engineers, Peter Peregrinus, 1990); Oskar Blumtritt, Hartmut Petzold, and William Aspray, eds., *Tracking the History of Radar* (Piscataway, N.J.: IEEE Center for the History of Electrical Engineering and the Deutsches Museum, 1994); Kern, "Die Enstehung des Radar Verfahrens."

SEVEN

GERMAN RADAR

$\mathcal{H}$ANS E. HOLLMANN, one of the several pioneers in the field of radar, not only was a leader in microwave technology but played an indirect role in the reemergence of semiconducting silicon into electronics.[1] Hollmann was an early radio enthusiast and microwave pioneer who worked at various posts in or near Berlin, Germany, until 1945. He then migrated to the United States with the so-called "paper clip" group of scientists and engineers who were invited by the federal government to aid in civilian or military research, as well as to avoid the relatively high probability that they would become Soviet scientist-captives if they remained in the Communist part of Germany, as seemed to be the only other choice for Hollmann.

He was not sympathetic to his own government after 1933 as he began to comprehend Hitler's political and military objectives. As a result, he decided to carry on his work in a private organization devoted to medical electronics, created with his brother, in order to avoid political involvement to the extent possible under the difficult circumstances that prevailed. Although he had opportunities to obtain university posts in Germany in 1935, he refused them since such an appointment at that late date would have required that he become a member of the National Socialist party.

THE INTERNATIONAL POLAR EXPEDITION, 1932–33

Hollmann was not only deeply involved in all aspects of electronic and electromagnetic research in the 1920s and 1930s, but became a principal member of the German team that explored the ionization layers of the atmosphere and related effects during the international polar scientific expedition of 1932–33. This work led him to appreciate the potential of radar (a word actually not coined until 1940). He had made prior arrangements to have a reasonably well equipped laboratory workshop available in Norway so that he could service and improve his equipment as the research progressed. As a result, he developed a system that generated pulses of microsecond duration, considerably shorter than anything previously accomplished. He was able not only to get good reflections from the ionosphere, but also to view features of the surrounding terrain on an electron oscilloscope.

THE FIRST PULSED RADAR

Upon returning home, Hollmann was immediately involved in a major way in the industrial development of what became a standard German pulsed radar system that operated at 2.6 meters (115 megahertz). He worked through a company bearing the acronym GEMA (Gesellschaft für Elektro-Akustic und Mechanischer Apparat).

On leaving the microwave field in the mid-1930s, he became divorced from political and military affairs and no longer held military security clearance. It also appears that the military avoided him since he was not sympathetic to the National Socialist party. Out of concern for fellow civilians in the event of war and air raids, he later worked privately with Manfred von Ardenne, who directed a private electronics laboratory, on the development of a microwave radar system that could detect aircraft at night and in clouds.

WARTIME POST

In 1942, Hollmann was made scientific head of a newly founded organization, the Research Society of Wireless and Sound Film Technology, a private entity funded by the film industry, which gave him a great deal of freedom to carry on research of his own choosing, provided he stayed within governmentally mandated boundaries. This excluded novel microwave research, for reasons of secrecy. It provided him with funds to support research groups involved in what were regarded as significant research programs, within the same governmental guidelines, in countries occupied by Germany. For example, he was able to provide support to the Kammerlingh Onnes Institute in Leyden, Holland, for research on the rate of development of photographic films at various temperatures.

It appears from comments made by Denis Robinson that Samuel A. Goudsmit, the scientific head of the Alsos Mission, which examined the German work on nuclear reactors during World War II, actually met Hollmann in Berlin during the late summer of 1945 at the suggestion of Robinson, who knew Goudsmit well from overlapping years at the Radiation Laboratory at the Massachusetts Institute of Technology.[2] Goudsmit found that most of the German technicians in the organization Hollmann and his brother had established were eventually drafted into the army or sent to military laboratories, thus bringing that enterprise to an end.

HOLLMANN'S BOOK ON HIGH-FREQUENCY TECHNOLOGY

Before turning away from microwave research as his central activity, Hollmann wrote a two-volume treatise on high-frequency technology, *Physik und Technik der Ultrakurzen Wellen* (Physics and technology of ultra-short waves).[3] This work, published by Springer in 1936, was not fully appreciated at once, but it became a classical reference text for the field prior to the outbreak of World War II. Hollmann took great pains to cover every aspect of the development of high-frequency technology, both recent and historical, and to provide many illustrations. Apart from the problem of language, the books were rarely found in engineering libraries in the United States for two reasons: Springer books were usually very expensive; and there was a tendency to boycott German products during the Hitler period. Actually, one of the two volumes was reproduced under the authority of the U.S. Custodian of Alien Property during World War II. (One of the pair of volumes currently available in the U.S. Library of Congress is such a duplicated version). The Stanford University engineering library also possesses copies, presumably as a result of the interest surrounding the klystron, which was developed and improved at Stanford in the 1930s.

MULTICAVITY MAGNETRONS

The importance and capabilities of the multicavity magnetron, which was being perfected as a result of work in England, the Netherlands, France, the Soviet Union, and eventually in the United States, were just beginning to be appreciated in the mid-1930s. As a result, Hollmann's work provided an ideal introduction to technology that proved to be important for the development of high-powered microwave radar during World War II. Hollmann placed major emphasis on the cylindrical split-anode magnetron, which he termed the *Habann tube*. He clearly had a great deal of experience using the tube.

CRYSTAL DETECTORS

For purposes of historical reconstruction, it is worth noting that in the second volume of Hollmann's treatise on the physics and technology of ultra-short waves, he makes the following statement under the section "Detector Reception" (*Detektorempfang*): "Using current technology, the simplest rectifier and wave indicator to employ at ultra high frequencies is a crystal detector."[4] He then goes on to describe uses of relatively primitive point-contact crystal detectors with which he has had experience, emphasizing their virtues and limitations. More specifically, he mentions pyrite crystals (FeS_2) with bronze or iron whiskers. He also emphasizes the benefits of employing a diode with a very fine cat-whisker to reduce capacitance, and of searching for a very sensitive region on the crystal. As emphasized earlier, there is no indication that he or his immediate colleagues actually used anything resembling a silicon (or germanium) point-contact rectifier as a heterodyne mixer for radar reception. That was to come later. Nevertheless, the attention Hollmann directed toward the use of point-contact crystal rectifiers of low capacity was without doubt the factor that stimulated Denis Robinson to propose more general uses of such units, as will be seen in chapter 10.

SILICON REEMERGES

In a remarkable paper published in the open literature in 1938, and most probably influenced by Hollmann's publications, Jürgen Rottgardt, who was employed at the Electrophysical Institute of the German Air Force Research Establishment in Berlin, presented the results of an extensive study of cat-whisker–crystal rectifying combinations for use in the microwave region. He concluded that "the combination silicon-tungsten is particularly favorable for the production of detectors to be used at very short wavelengths."[5] Rottgardt worked in the wavelength range from 50 to 1.4 centimeters. No mention is made of the source of silicon, whether of metallurgical or reagent quality. This was clearly one of the landmark studies, although it appears to have been buried in the uncited literature until discovered recently by Berthold Bosch of the Ruhr University at Bochum. Rottgardt's paper was followed shortly thereafter by another, which focuses further on silicon-tungsten, prepared by a coworker, H. Klumb.[6] Again, the source of the silicon used is not given.

Why was the important research of Rottgardt and Klumb on silicon diodes treated so casually by those in charge of German military radar, inasmuch as they had subjected the second volume of Hollmann's book on high-frequency technology to strict censorship on security grounds? As Louis Brown of the Carnegie Institution in Washington, D.C., believes, the answer is probably a simple one: Those in charge of the development of German military radar in the late 1930s looked upon the crystal diode as a testing or measuring device and not as an essential circuit element for radar.[7] Unlike the French and British, they did not appreciate and seek out the potentialities for developing high-powered transmitters in the centimeter range for which such diodes would become essential components as heterodyne mixers. Apparently, no one with Hollmann's forward-looking perspective in the microwave field was involved in the planning of German military radar at that stage, or, if there were such individuals, they were not encouraged to seek more powerful sources in the centimeter range.

GERMAN RADAR: LATE 1930S TO 1945

German radar technology subsequently moved ahead at a relatively slow pace until very late in World War II when the full significance of developments in the United Kingdom and the United States came to light as a result of studies of radar equipment found in downed Allied aircraft. Prior to that time, however, some equipment using 50-centimeter wavelengths (600 megahertz) was put in operation for directing gunfire. The effort to develop operating equipment at shorter wavelengths came too late to be fully useful during the war.

Between the late 1930s and 1945 most of the research on microwaves and the associated generating and detecting systems in Germany was carried out either at industrial organizations such as the Siemens Electric Company and Telefunken or within military laboratories, particularly those of the German air force. Young, bright scientists and engineers were not present in large numbers because Hitler had insisted that all who could must serve in uniform in the regular armed services—a very major error in judgment on the part of the leadership. Nevertheless, significant developments were carried out during the war by individuals such as W. Bül, Herbert Mataré, Karl Seiler, and Heinrich Welker with limited staff.[8] For example, Welker, who, later became well known for his research on semiconductor compounds of elements from the third and fifth column of the periodic chart, such as gallium arsenide, was making excellent detectors and mixers of germanium as early as 1941 and 1942. Karl Seiler and his group discovered the gaps of electron levels in silicon and germanium in 1942 but were not permitted to mention them explicitly in their papers since the venerable physicist B. Gudden, a pioneer in solid-state electronic devices, insisted that the two crystals were metals.

MILITARY ATTITUDES

In an essay written in 1994 to commemorate the sixty-fifth birthday of a colleague, Professor Berthold Bosch mentions that, initially, the installation of radar in German fighter planes was strongly opposed by Marshall Hermann Göring, who was viewed as the father of the German air force, because he believed that the intrinsic capabilities of the fighter pilots ought to be superior to electronic observation systems. Göring insisted that there be "no movies!"[9] He had been a celebrated, open-cockpit fighter pilot in World War I and had little appreciation of the scale of advances of technology or what it meant for modern warfare. Changing circumstances, late in the war, altered his view. The abrupt change in outlook started in February 1943 when a radar system containing an advanced ten-centimeter magnetron was recovered from a fallen Allied plane. Göring ordered a rapid examination of its capabilities and, at the same time, requested a recall of some fifteen hundred technically trained soldiers who had been placed in combat divisions. By summer, a primitive copy had been tested and the capabilities of such equipment for the bombing of cities at night and in bad weather, and against submarines, was fully appreciated. In the haste to manufacture comparable equipment, use was made of concentration camp prisoners, a matter that was later made the subject of charges in the Nuremberg trials.

In 1940 Manfred von Ardenne had visited Göring to express the interest he, with Hans Hollmann's help, had developed in producing a microwave radar that might be used defensively in case of enemy bombing raids. He was waved off with the comment that the war was essentially over. In 1943, by which time the bombing the Germans had feared was heavy, von Ardenne approached an admiral in charge of naval electronics, an old family friend, with a similar offer. In this case, much to his dismay, he was again brusquely turned away. Apparently, an association with independent civilian scientists represented a hazard to officials by that time.

In this connection, a distinguished German physicist, Karl Ramsauer, visited Propaganda Minister Paul Göbbels in May 1943 and had the courage to say to him that Germany was being overwhelmed as a result of the very successful use the Western Allies were making of their competence in physics, chemistry, and related technology. In contrast, the German government had not provided any sensible leadership for the use of its own scientific talent for the past ten years and was now paying the price for it. To quote what appeared in Göbbels diary, Ramsauer said to him: "You cannot expect to put an airhead *(Hohlkopf)* on the top of a ministry and expect research to provide sensational results."[10]

Hans E. Hollmann, a pioneer in research in the field of high-frequency electromagnetic radiation. He participated in 1934 and 1935 in the development of what was probably the first effective pulsed radar system. It operated at 2.6 meters. Hollmann migrated to the United States in 1945. (From the discontinued journal *Hochfrequenztechnik und Elektroakustik* 68, no. 5 [1959]: 141.)

Samuel A. Goudsmit, the scientific head of the Alsos Mission, which was sent to Europe in 1944 to determine how successful the German scientists had been in exploiting nuclear fission. He met Hans Hollmann during August 1945 and learned aspects of his wartime experiences. (Courtesy of the American Institute of Physics Emilio Segrè Visual Archives, Physics Today Collection.)

Physik und Technik der
ultrakurzen Wellen

Von

H. E. Hollmann
Dr.-Ing.

Zweiter Band
Die ultrakurzen Wellen
in der Technik

Mit 283 Textabbildungen

Berlin
Verlag von Julius Springer
1936

The title page of the second volume of Hollmann's treatise on high-frequency radiation, published in 1936. The books were far in advance of anything else of the kind at the time. They caused some degree of consternation among individuals concerned with the rapid pace of German rearmament. (Courtesy of Franco Bassani and Gianfranco Chiarotti.)

Manfred von Ardenne, who operated a highly productive, private electronics research laboratory in Berlin. He played a leading role in the early development of radio, television, and the electron microscope. (Courtesy of von Ardenne.)

Carl Ramsauer, the scientist who, in 1943, had the courage to criticize the National Socialist government in Germany for its lack of understanding of science and the scientific community. (Courtesy of the Deutsches Museum, Munich.)

NOTES

1. For other discussion about Hollmann, see chap. 3.

2. Denis M. Robinson, *Proceedings of the American Philosophical Society* 127 (1983): 26. A vain search of the Goudsmit files in the Archives of the American Institute of Physics indicates that Goudsmit probably did not preserve a written record of his encounter with Hollmann in Berlin.

3. H. E. Hollmann, *Physik und Technik der Ultrakurzen Wellen,* 2 vols. (Berlin: Springer, 1936).

4. Ibid., 2:2.

5. J. Rottgardt, *Zeitschrift f. Technische Physik* 19 (1938): 262.

6. H. Klumb, *Physikalische Zeitschrift* 40 (1939): 640; H. Klumb and B. Koch, *Die Naturwissenschaften* 27 (1939): 547.

7. Louis Brown, *Technical and Military Imperatives: A Radar History of World War II,* forthcoming from the Naval Institute Press, published in cooperation with the IEEE.

8. H. F. Mataré has prepared an account of his postwar research and development in the field of semiconductor electronics for a special issue of the *Proceedings of the IEEE* to appear early in 1998. It has the title "Lesser-Known History of the Crystal Amplifier" and deals with his activities in both Europe and the United States. It contains a brief account of the development of germanium triodes in 1948 by a French team in which he participated (see page 174 of this volume). The first open technical report of this work, Mataré's "The Three-Electrode Crystal (Transistor)," appeared in the German journal *The Electron in Science and Technology* 3, no. 7 (1949): 255. We are grateful to Mataré for the privilege of seeing a prepublication draft of his forthcoming paper, as well as other technical and biographical material. As early as 1951 he published a book, *Empfangsprobleme im Ultrahochfrequenz Gebiet* (Reception problems in the realm of high frequencies) (Munich: R. Oldenbourg Press), that devoted much attention to the use of semiconductor devices in the microwave region.

9. The comment by Hermann Göring appears in the unpublished essay by Berthold Bosch, "Der Werdegang des Transistors 1929–1994, Bekanntes und Weniger Bekanntes," an address given on Nov. 17, 1994.

10. The quotation from the statement made by Karl Ramsauer appears in Ulrich Kern, "Die Enstehung des Radar Verfahrens: Zur Geschichte der Radar Technik bis 1945" (Thesis, University of Stuttgart, 1984), p. 246.

EIGHT

FRENCH RADAR

$\mathcal{F}$RENCH INVOLVEMENT IN RADAR prior to 1939 parallels developments in Germany,[1] although the initial interest during the early part of the
1930s focused on continuous wave (CW) systems used to create something
in the nature of an electromagnetic fence or cordon that would detect tres-
passing aircraft. This point merits emphasis, particularly since the British
started their radar detection network with a similar continuous wave barrier
until it became clear that a pulsed system would be superior. The leading
individuals in the initial work in France were Maurice Ponte, Henri Gutton,
Sylvain Berline, and M. Hugon, who followed up on the by then commonly
observed reflections of radio waves from aircraft first reported by R. Mesny
and Pierre David in 1931. This eventually led to work on pulsed systems
at decimeter (sixteen cm) wavelengths with a major focus on magnetron
research.

In his book *Souvenirs de longue vie,* Emile Girardeau emphasizes that
the French teams were working with radiation in the submeter range as
early as 1934 and 1935 and states, with considerable justification, that the
French contribution between that period and May 1940 has been inad-
equately appreciated.

To quote one of the leading French scientist-engineers, Pierre Aigrain:

> My own feeling is that the most important contributions of French
> scientists and engineers to the future developments of radar were those
> of Ponte, Gutton, Berline, and Hugon on the early development of mag-
> netrons. They moved from the unpredictable Hull magnetron (with a
> simple cylindrical anode) to the split anode type, then to the interdigital
> type, and in 1940 finally built a cavity magnetron, an obvious (*a poste-
> riori*) step forward from the interdigital anode. Ponte later became presi-
> dent of C.S.F., but his most important contribution was to bring their
> prototype to England at the end of May 1940, so the Germans would not

get hold of it, and the British could use it as a starting point for future improvements.

In fact, as early as 1935, a simple, hard to operate, low performance real radar was installed on the famous liner the *Normandie* as an iceberg detector. It used, I think, an interdigital anode magnetron operating at a wavelength of 16 centimeters. (Photo page 166 of Antebi's book.)[2]

It is particularly noteworthy that some developments on magnetrons, particularly with respect to large-area oxide-coated cathodes, carried out in France and transferred to England just prior to the fall of France, played a very significant role in advancing the art of high-powered, centimeter-wavelength, multicavity magnetrons.[3] The large-area cathodes could be pulsed for high emission currents. Moreover, secondary emission from the cathode played an important role in achieving high levels of radiated power. Ponte's journey to England with the precious French magnetron took place just in the nick of time, on May 8, 1940, immediately before the fall of Paris.

Pierre David offered praise to the British for their approach to the problems, as well as for their dedication and achievement in the field of microwave radar.[4] Among their several major achievements, he listed "*les detecteurs à cristal à grande sensibilité.*" This comment suggests that the French workers were unable to achieve comparable sensitivity with the crystal detectors they used prior to the spring of 1940. It probably explains the fact that the French government focused special attention on the development of germanium crystal diodes immediately after World War II. In the process, the team independently invented a semiconductor triode (see section starting on page 174).

In his biographical account of Hans Hollmann's professional career, H. Frühauf states that much of the early work on the cavity magnetron was carried out at the Philips Eindhoven laboratories in the Netherlands and that the French telecommunications specialists formed a close relationship with Philips, leading to French patents in the field.[5] Hollmann followed this work through a link with a private German company other than GEMA mentioned in the previous chapter.

Maurice Ponte, one of the leaders
in the development of the French
version of the cavity magnetron.
He took the advanced French
model to England just before the
fall of France in early May 1940.
(Courtesy of Pierre Aigrain and
the Archives of the Academie des
Sciences, Paris. The photograph
has been reproduced by Jean-Loup
Charmet.)

NOTES

1. See Ulrich Kern, "Die Enstehung des Radar Verfahrens: Zur Geschichte der Radar Technik bis 1945" (Thesis, University of Stuttgart, 1984); Henry E. Guerlac, *Radar in World War II,* 2 vols. (New York: American Institute of Physics, 1987). See also Emile F. E. Girardeau, *Souvenirs de longue vie* (Paris: Berger-Levrault, 1968); Elizabeth Antébi, *The Electronic Epoch,* (New York: Van Nostrand, 1982), p. 167.

2. Personal correspondence, Pierre Aigrain to Frederick Seitz, July 6, 1994.

3. See E. B. Callick, *Metres to Microwaves* (London: Institution of Electrical Engineers, Peter Peregrinus, 1990).

4. Pierre David, "Quelque commentaires sur l'histoire du radar," *L'Onde electrique,* p. 15; volume and year unknown, but probably in the early 1950s. We obtained a photocopy of this undated item from Pierre Aigrain.

5. See H. Frühauf, biography of Hollmann in *Hochfrequenztechnik und Elektroakustik* 68, no. 5 (1959): 141.

NINE

SOVIET RADAR

$\mathcal{I}$NDIVIDUALS from the United States who visited the wartime Soviet Union in connection with the delivery of military equipment reported that the radar units sent were assembled and put into operation with lightning speed by highly expert Soviet teams. This indicated clearly that there was great interest in the technology and that innovation and production were limited only by wartime shortages and perhaps by Stalin's purges, which had the effect of stifling innovation by both the military and civilians.

THE MULTICAVITY MAGNETRON

A team under N. F. Alekseev and D. D. Malyarov published a remarkable paper in the open literature in the 1940s. It described an important research program on the cavity magnetron that had been started in the mid-1930s and was clearly based upon an ongoing program in the development of radar.[1] The scientists had created a relatively powerful water-cooled unit and demonstrated the characteristics and capabilities of the device. They achieved an emission level of three hundred watts at a wavelength of nine centimeters. Their paper, which appeared in a Russian-language technical journal, was picked up by an electronics engineer at the General Electric Company who had it translated into English and then published in 1944 in an American journal. The presentation implies that the investigators used a large-area oxide-coated cathode since the relative sizes of the central cavity and the cathode were adjusted to match when setting the operating conditions. While the magnetrons developed by the Western Allies eventually achieved more than ten times the power of the Soviet one, the result was remarkable for 1940, under the circumstances.

The Soviet paper appeared precisely at the time when the British workers regarded any information concerning the effectiveness of the cavity magnetron to be at the very pinnacle of their list of secrets. If they were aware of the Soviet program, they obviously made no open mention of it.

Since the paper appeared in the open literature, it could have guided the German technical staff into work on cavity magnetrons had there been any incentive on the part of their leadership to extend radar research to the centimeter range in 1940. The German and Soviet leaders had a formal cooperative alliance at the time the paper was published, although it is difficult to know how effective the agreement actually was.

It is interesting to observe that in a paper published by H. Klumb in 1940, the author lamented the fact that the intensity of radiation produced by split anode magnetrons and similar devices available to him falls to very low levels in the centimeter range of wavelength.[2] Apparently the German workers were not familiar with the Soviet work at that time and almost certainly failed to follow the Soviet technical literature. As mentioned earlier, they became familiar with the great virtues of the cavity magnetron only after retrieving one from a fallen Allied bomber in 1943.

CORRESPONDENCE WITH ACADEMICIAN ZHORES ALFEROV

Zhores I. Alferov, director of the A. F. Ioffe Physico-Technical Institute in St. Petersburg and vice president of the Russian Academy of Sciences, reveals much about the status of microwave developments in the Soviet Union prior to World War II. In one of his letters, he wrote:

> I was educated for vacuum tube electronics, but started to do semiconductor research in 1951 when I was a student. I remember from my student years how deeply I was impressed by D. Malyarov and N. Alekseev's work on high-power many-cavity magnetron. The idea of this work belonged to Professor M. A. Bontch-Bruevitch who was a pioneer in radioelectronics in general in our country and D. Malyarov was his disciple. They carried out research at the NII-9 (Research Institute No. 9), a secret institute created in 1935 in association with the Leningrad Electrophysical Institute. The latter was founded by Academician A. A. Tchernyshev, A. F. Ioffe's deputy, and branched out of our Physico-Technical Institute in 1931. The first pulsed radar system was created at the Ioffe Institute by Professors D. Rozhansky and Yu. B. Kobzarev in the middle of the 1930s. They built the first radar station at Toksovo (suburb area of Leningrad), and this station operated successfully during the siege of Leningrad.[3]

Alferov and three colleagues also provided a historically valuable paper covering the history of semiconductor research at the Leningrad Physico-

Technical Institute up to 1969.[4] The institute was clearly a fountainhead of creativity in electronics and related areas of research and development. Although such a center was envisioned in pre-Revolutionary days by leading scientists, particularly M. I. Nemenov, it did not receive appropriate governmental support until after the Revolution, when Nemenov was encouraged to make it a truly productive center.

Nemenov's personal interest lay in the medical applications of X rays. As a result, he selected A. F. Ioffe, a widely experienced physicist who had studied under Röntgen, to provide background support. Eventually, the activities of the original institute were split so as to create three institutes, and Ioffe became an independent head in laboratories devoted to problems in physics and chemistry, both exploratory and applied. The paper makes it clear that Ioffe's influence led to extensions of activity in many fields of electronics, including radar.

Although much work was carried out on semiconductors such as selenium and cuprous oxide soon after Ioffe arrived, the interest in germanium and silicon and the three-five compounds such as gallium arsenide started after World War II, at which time Ioffe and his colleagues became major participants in the field on the worldwide scientific stage.

The Soviet work on radar grew out of a division of Ioffe's institute that had been concerned with various aspects of vacuum tube electronics, as well as the transmission and reception of radio waves of various frequencies. The division began its life in the 1920s when, among other things, it manufactured special tubes as a national service.

(Facing page, bottom) The Soviet team that carried out research and development on advanced radar systems in the 1930s and during World War II. The group developed a remarkably powerful cavity magnetron, which was described in a paper in the open literature published in 1940. *Left to right: (front)* D. D. Malyarov, A. E. Sysant, V. N. Mydrogin; *(rear)* A. Z. Fradin, M. D. Gyrevich, N. F. Alekseev. The paper mentioned was published by Malyarov and Alekseev. (Courtesy of Zhores Alferov and Robert Cahn, with photographic enhancement by Freelance Photographic Services, Somersham Huntingdon, England.)

A. F. Ioffe, the head of the Physico-Technical Institute in Leningrad. He was a great scientific leader who encouraged the advancement of many applied fields, including electronics and solid-state chemistry and physics. (Courtesy the American Institute of Physics Emilio Segrè Visual Archives.)

NOTES

1. N. F. Alekseev and D. D. Malyarov, *Journal of Technical Physics* 10 (1940): 1297 (Russian article); published in English in *Proceedings of the Institution of Radio Engineers* 32 (1944): 136.

2. H. Klumb, *Zeitschrift f. Physik* 115 (1940): 321.

3. Letter, Z. Alferov to Frederick Seitz, Apr. 12, 1996.

4. Zh. I. Alferov, V. I. Ivanov-Omskii, L. G. Paritskii, and V. Ya. Frenkel, "Investigations of Semiconductors at the Physico-Technical Institute" (in English), *Soviet Physics—Semiconductors*, vol. 2 (New York: American Institute of Physics, 1969), p. 1169.

TEN

BRITISH RADAR

$\mathcal{A}$DVANCED MICROWAVE RESEARCH tapered off substantially in Germany after the development of the primary radar involved in the armed services.[1] Moreover, in the United States the topic was treated mainly as a field for basic investigation by universities and the communication industry. (For example, W. W. Hansen of Stanford University, working in close cooperation with two brothers, Russell and Sigurd Varian, and with D. L. Webster, developed the klystron initially for possible use in nuclear accelerators.) The British government, however, fully aware of the great potential threat of German rearmament, particularly the rapid development of German bomber planes, decided in 1937 that its fighter planes needed to be equipped with radar and that much of its available technical strength should be drawn into the field. For resolution and compactness, it was decided that the systems to be developed should be based on pulsed radiation in the centimeter range of wavelengths.

THE MAGNETRON

The British pushed ahead in great secrecy with the development of small, powerful multicavity magnetrons. Through special, highly inventive design, and significant help from French technology, they eventually achieved much higher outputs of pulsed power at short wavelengths than those reached by investigators in other countries who were also working with magnetrons. The responsibility for carrying out these developments was in the hands of a specially created organization, ultimately designated the Telecommunications Research Establishment (TRE). It was formed under the guidance of a high-level advisory committee established in 1935 and chaired by Henry Tizard.[2] It contained a substantial component of the leading scientific and engineering talent in the country.

In the days of need, in fact, the scientific and engineering leaders in Britain joined forces in an extraordinary way under government sponsorship to develop critically useful radar. They were prepared to create and borrow as required to achieve results as rapidly as possible. While those involved preserved a high degree of individual freedom in the approach to their work, the urgency of the situation they faced inevitably led to much cross reaction and coordination. A list of those involved in the program represents a catalog of some of the most eminent leaders of government, industrial, and academic science and engineering.

To gain a reasonable view of the situation, one can hardly do better than to read chapter 6 of Henry Guerlac's history of radar.[3] In addition to Tizard, the account includes such outstanding figures as Patrick Blackett, Henry A. H. Boot, Edward G. Bowen, John D. Cockcroft, Archibald V. Hill, Wilfred Bennet Lewis, Frederick A. Lindemann, Mark L. E. Oliphant, A. P. Rowe, John T. Randall, Robert Watson-Watt, and Arnold F. Wilkins. While some, such as Lindemann, who was adviser to Prime Minister Churchill, and Watson-Watt, who initially was superintendent of the radio division of the National Physical Laboratory, held official government positions, most came out of research laboratories and had gained substantial recognition through earlier work in other fields.

Apparently the first moderately comprehensible theory of the operation of a magnetron that involved treating the electron space charge as a coherent entity was developed in Holland by the Dutch engineer K. Posthumus of Philips Eindhoven, which carried out much of the early research on the device. Issues related to the collective behavior of electrons in the circulating space charge were treated sparingly in this epoch.

DENIS ROBINSON AND H. W. B. SKINNER

There were limited large concentrations of technical resources in the United Kingdom. However, Henry Tizard, who had gained much experience with operations research during World War I, had learned that specific projects of considerable importance could, with benefit, be farmed out to individuals or groups of investigators in small companies or academic laboratories who had special talents. In this way, Denis Robinson, an English electrical engineer who had been working in a private company in England in 1939 (Scophony Television Laboratory) carrying out audio and video electronic research and development for commercial production, found himself sent by his government to a laboratory in Dundee, Scotland, that was devoted to secret work. There he was asked to "develop a detector circuit for reception from a potentially available ten centimeter microwave source." He thus became involved in a satellite operation of the Telecommunications Research Establishment that was eventually centered at Great Malvern in the Cotswolds. As Robinson recalled some fifty years later, in 1991:

> I was pretty impressed by this demand placed on me to get a receiver, because the coils got smaller, and everything had now disappeared into the only tubes we had. So I knew that we had to have a totally new look.
>
> I went to the library and immersed myself there, and fortunately, found a little book called Hochfrequenztechnik (in German), produced by the wonderful Springer publishing company. It was by a man called Thoma, who was a professor at one of the lesser-known universities in the south of Germany. He had not the slightest idea that it was going to be used for any war or other purpose, but he studied the whole thing and said, "How do we make the next advance?" You've done the work on Hertz, so you know. He understood all that stuff, and like any good German professor, he had to start with Ohm's Law [chuckling] in the book. He went through, and he gradually rejected every receiver that had a valve or tube or something like that. He said, ". . . the self-capacity there is more than we can use, so the only thing. . . ." Finally, he came to the conclusion, ". . . the only thing we can use for a receiver is the crystal and cat's whisker." Well, I was delighted because that's what I'd used ten years previously. So I went back to W. B. Lewis and said, "Look, this is it." He could read German enough to read these pages that I'd marked for him, and he said, "Let's start!"[4]

In other words, Robinson decided that he could not rely on using any potentially available vacuum tube as a mixer in a heterodyne circuit that would be used to reduce the frequency of the return signal from a radar pulse to a reasonable practical range for amplification in an amplifier operating at a much lower frequency. He was, however, not only brilliant but had what is termed a "well-prepared mind." Something he saw in the German literature made it clear to him that his best hope in the short—and probably in the long—run was to use a rectifying cat-whisker crystal diode as the nonlinear element in the heterodyne circuit.

Denis Robinson is not mentioned in Callick's book Metres to Microwaves,[5] which focuses primarily on the activities of investigators who contributed more or less full time in one way or another to the work of the Telecommunications Research Establishment between 1937 and the end of

World War II. Perhaps this omission is due to the fact that Robinson was assigned to the Radiation Laboratory of the Massachusetts Institute of Technology in the summer of 1941 and completed his period of wartime service there as a highly valued and well-recognized member of the staff. Moreover, Callick credits M. L. E. Oliphant with discovering the importance of crystal diodes in the centimeter range. This is, perhaps, not surprising since Robinson was junior to Oliphant and, in a sense, he reported to him. H. W. B. Skinner, however, is given appropriate credit for his work on diodes and other matters.

For collaboration, Robinson turned to Skinner, who had been at the University of Bristol, an international center for solid-state research under the leadership of N. F. Mott, and was now at the TRE.[6] They soon found, after trying a number of combinations of crystals and whiskers with materials available to them, that metallurgical-grade silicon, used in combination with tungsten wires, provided the only satisfactory solution. Or, as Skinner, who then took over the problem of early fabrication, stated in notes written in 1948: "Other substances than silicon were tried without very much success."[7] It is possible that Skinner was familiar with the important works of Rottgardt and Klumb. Rottgardt's paper, however, is not cited in any issue of *Science Abstracts* of the period, so it could easily have been overlooked.

The significance of Skinner's role in the practical development of point-contact silicon rectifiers, after Robinson's decision that it was necessary to use crystal diodes, is indicated clearly in a letter written by W. E. Burcham of the University of Birmingham:

> The breakthrough from a practical point of view came on 16 July 1940 when H. W. B. Skinner found that silicon could be brazed onto a tungsten rod so that both the silicon and tungsten wire point contact could be sealed into a glass envelope which was subsequently filled with a viscous liquid to damp vibration. These glass-enclosed crystals were used in our flight trials of the radar system until overtaken by the capsule units produced commercially by the BTH (British Thomson-Houston) Company. The BTH development was encouraged by M. L. E. Oliphant, whose group at Birmingham had assembled a klystron-driven centimeter radar.[8]

At that time, Robinson had become deeply involved as administrator of another urgent major radar program involving fifty-centimeter microwave radiation.

As will be discussed in chapter 13, an independent discovery of the virtues of the tungsten-silicon diode was made by R. S. Ohl at the Bell Telephone Laboratories in connection with microwave research under way there.

One may wonder about the special attributes of the silicon-tungsten combination that permitted it to be discovered independently by a number of investigators searching for a good rectifying diode in the microwave range. Probably several factors are involved, including the relative chemical, thermal, and mechanical stability of the components. Also, it is likely that the hard tungsten whisker, when suitably pressed against the silicon by mechanical means, breaks through any oxide layers present and makes good physical and chemical contact with the silicon.

It was found, in the course of the wartime research with silicon diodes, that tungsten forms a better rectifier with p-type than with n-type silicon, whereas the converse was true for germanium.[9] When n-type germanium (see chapter 4 for this terminology) is brought to electrical equilibrium with tungsten, it contributes electrons to tungsten (see illustrations on page 82) and must, for good rectification, form a substantial barrier between the two. In other words, prior to the equilibrium adjustment of charge, the pair acts as if the top of the occupied levels of the tungsten lies somewhere near the bottom of the relatively small gap, of the order of 0.7 eV, between occupied and vacant bands of germanium. The same line of reasoning would suggest that, prior to contact, the top of the occupied levels in tungsten lies near the upper part of the band gap in silicon. This might appear, superficially at least, to represent an interesting, fortuitous form of relative potential balance that, among other things, makes it possible to have good contact rectification, of opposite sign, for the two semiconductors. In this connection, it may be noted that iron, which has a work function close to that of tungsten, was commonly used as a cat whisker in combination with silicon in the days of wireless telegraphy.

Actually, as was first emphasized by J. Bardeen and studied more fully by C. A. Mead and W. G. Spitzer, who employed a variety of metals in conjunction with semiconductors having the diamond structure, the development of an electric dipole layer associated with trapping levels (see chapter 14) at the contact interface between the metal and semiconductor plays a critical role in determining the relative positions of the levels in the metal-semiconductor combinations.[10] The magnitude of the dipole field may in turn be influenced by the details of the procedure followed when fabricating diodes. Thus the values of the dipole layers that actually occur in the diodes may be determined in part by experience that is gained in manipulating the contact between cat whisker and semiconductor in the course of manufacturing the devices.

POSTWAR CAREERS

Robinson returned only briefly to England at the end of World War II, prior to taking up residence in the United States as president of the High Voltage Engineering Company in Massachusetts, which he founded with John Trump and Robert van de Graaff. He remained a resident of that state until his death in 1994. He was a much-admired, soft-spoken leader of those who worked with him.

Skinner left radar work in England in 1943 to join the Manhattan District Laboratory of E. O. Lawrence in Berkeley, which was engaged in isotope separation by mass spectroscopic means. Skinner returned to England at the end of the war to head the physics division at the Atomic Research Establishment at Harwell where he played a major role in establishing programs and standards at the laboratory. He left Harwell in 1949 to become professor of physics at the University of Liverpool, but he died prematurely in 1960 at sixty years of age. His frustrations with wartime research on radar are expressed in a poem that he wrote in 1941. One stanza reads:

> And so alone,
> we, fighting every inch of the way,
> against those ingrained elephants of inertia,
> against the prejudice and the hardened pride
> of self-established, self-supporting systems,
> we fought (through forests, thick with self-satisfaction)
> to shorter electromagnetic wavelengths.[11]

Evidently Skinner encountered roadblocks to his work that were raised by specialists of the vacuum tube era. It seems to be a general rule that a new branch of technology, like a new field of science, must struggle to "earn" its place in the spectrum by proving to be indispensable. Moreover an entirely new generation, relatively free of older forms of mindset, normally assumes the role of leadership in the emerging field.

INDUSTRIAL INVOLVEMENT

While some diodes were made at the Telecommunications Research Establishment, as indicated in Burcham's letter, the main problem of manufacture had been turned over to the British electrical industry by early 1941, primarily to British Thomson-Houston (BTH) and the (British) General Electric Company (GEC), with guidance from TRE. Actually, GEC appears to have carried out some research with the silicon-tungsten combination on its own early in 1940, perhaps as a consequence of the early results obtained by Robinson and Skinner or possibly by the German workers. In any event, practical tests, sparked by the initiatives of Robinson and Skinner, led back to crystal rectifier technology that had been all but abandoned fifteen or so years earlier and with which Robinson had experience from early amateur radio or experimental work. The silicon diode was reborn again in time of need for a new use.

BRITISH SOURCES

While there does not appear to be a single source that provides a completely detailed picture of the British development of silicon diodes comparable to that appearing in volume 15 of the Radiation Laboratories Series, prepared by the team at the Radiation Laboratory in charge of the diode program (see next chapter), some excellent summaries of the chemistry and physics involved were provided by B. Bleaney and his colleagues in the immediate postwar period.[12] Progress was undoubtedly conditioned by research at the industrial organizations, which eventually focused on a form of available metallurgical-grade silicon doped with additives of aluminum. The electric companies succeeded in producing effective units notwithstanding the inherent variability of the available grade of silicon. That the units produced in Britain actually possessed considerable variability in behavior is demonstrated by the fact that tests were made on individual diodes and those that behaved particularly well in tests were distinguished by being marked with a red dot.

THOMA OR HOLLMANN?

Although Robinson credits the source of his inspiration to a book written by Alfred Thoma,[13] an investigation by John Bryant of the University of Michigan into the records of the historian Henry Guerlac shows that in 1944 Robinson actually credited it to something he found in the books of H. E. Hollmann (see chapter 7), which seems far more reasonable. Robinson's lapse in memory after fifty years is understandable. Moreover, Alfred Thoma, who was primarily interested in mathematical matters related to physics, was a close, younger associate of Hollmann and had published several papers with him, as well as a book on his own.[14] The book dealt with mathematical topics related to technology and physics. In it, Thoma refers to Hollmann's two books. Thoma's book, however, was substantial, containing 684 pages. Thoma remained active in microwave research and development during the war, working with GEMA. He accepted a post as head of a secondary school (Oberschule) in the town of Fulda in the Hessian town of what was then West Germany in 1953. It seems probable that he started teaching at a secondary school in Berlin, at least part time, as early as 1939 and that he had a parallel academic career.

Henry Tizard, a scientist and veteran of operations research in World Wars I and II. In 1935 he was made chairman of a committee that developed the plans that led to the creation of the Telecommunications Research Establishment and the subsequent development of advanced radar systems. (Courtesy of the Royal Society of London and the Godfrey Argent Studio, London.)

Denis M. Robinson, the perceptive English electronics engineer who participated in the development of commercial television in Britain in the 1930s, prior to the outbreak of war. He joined the staff of the Telecommunications Research Establishment for radar research, first as an "outside" member. The authors are indebted to Professor R. Everson of the Rockefeller University for computer replication of this photograph, which is based on a World War II original appearing in *Five Years at the Radiation Laboratory* (Cambridge, Mass.: Massachusetts Institute of Technology, 1946; reprinted by IEEE for the International Microwave Symposium, 1991).

Mark L. E. Oliphant *(center)*, the Australian physicist who took a leading position in the Telecommunications Research Establishment in the early days of World War II. The other individuals in the photograph are G. H. Dieke *(left)* and E. R. Rasmussen. (Courtesy of the American Institute of Physics Emilio Segrè Visual Archives. Photo by Paul Ehrenfest Jr.)

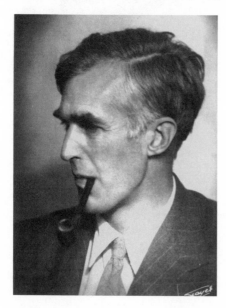

H. W. B. Skinner, to whom Robinson turned for cooperation. Skinner took over the crystal diode research program when Robinson became involved in a more immediately pressing radar problem. (Photo by Fayer of London. Courtesy of Robert Cahn, the Photographic Studio, and the Royal Society of London.)

NOTES

1. See Ulrich Kern, "Die Enstehung des Radar Verfahrens: Zur Geschichte der Radar Technik bis 1945" (Thesis, University of Stuttgart, 1984). See also Henry E. Guerlac, *Radar in World War II*, 2 vols. (New York: American Institute of Physics, 1987); and idem, *Journal of the Franklin Institute* 250 (1950): 285.

2. A biography of Henry Tizard appears in *The Biographical Memoirs of the Fellows of the Royal Society*, vol. 7 (1961), p. 313. See also the obituary by Patrick M. S. Blackett in *Nature*, Mar. 5, 1960.

3. See Guerlac, *Radar in World War II*.

4. Recorded interview with Denis Robinson carried out by John H. Bryant and included in his book *Radlab: Oral Histories Documenting World War II Activities at the MIT Radiation Laboratory* (Piscataway, N.J.: IEEE Center for the History of Electrical Engineering, 1993), p. 281. Denis Robinson's son, Harald Robinson of Arlington, Massachusetts, has two videotaped interviews with his father made by the High Voltage Engineering Company in the early 1990s. The authors are grateful for the privilege of viewing them, as well as for much additional information regarding his father.

5. See E. B. Callick, *Metres to Microwaves* (London: Institution of Electrical Engineers, Peter Peregrinus, 1990).

6. A biography of H. W. B. Skinner prepared by Harry Jones appears in *The Biographical Memoirs of the Fellows of the Royal Society*, vol. 6 (1960), p. 259.

7. Skinner, quoted in Callick, *Metres to Microwaves*, p. 91.

8. Letter, W. E. Burcham to F. Seitz, May 1995.

9. See H. C. Torrey and C. A. Whitmer, *Crystal Rectifiers*, The Radiation Laboratory Series, vol. 15 (New York: McGraw-Hill, 1948).

10. See J. Bardeen, *Physical Review* 71 (1947): 717; C. A. Mead and W. G. Spitzer, "Fermi Level Position at Metal-Semiconductor Interfaces," *Physical Review* 134 (1964): A713. We are indebted to H. R. Huff for calling our attention to the second paper.

11. Poem by Skinner quoted in Jones, *Biographical Memoirs*, p. 259.

12. B. Bleaney, "The Crystal Valve," *Journal of the Institution of Radio Engineers* 93, part 3A (1946): 184; B. Bleaney, J. W. Ryde, and T. H. Kinman, "Crystal Valves," ibid. 93, part 3 (1946): 847; ibid. 94, part 3 (1947): 339.

13. A brief account of Alfred Thoma's career up to 1961 appears in J. C. Poggendorff, *Biographisch-Literarisches Handwörterbuch der Exak. Wissenschaften*, vol. 7A, part 4 (S–Z) (Berlin: Akademie-Verlag, 1961).

14. See A. Thoma, *Die Differentialgleichungen d. Technik und Physik* (Leipzig: v. W. Hort, 1939).

ELEVEN

THE UNITED STATES:

The Radiation Laboratory

*S*CIENTISTS AND ENGINEERS in the United States were not unified in their approach either to radar or to the exploration of the microwave region of the spectrum prior to the outbreak of war in Europe in September 1939.[1] On the civilian side, some individuals were exploring frequencies in the gigahertz range and above for their own purposes. This led to the invention and development of devices such as the klystron and traveling wave tubes. As mentioned in the previous chapter, W. W. Hansen at Stanford University developed the klystron with the intention of using it to accelerate particles to high energies in a linear system—a linac. Others, such as the Varian brothers, who had been students at Stanford and were starting an electronics company in the area that bears their name, and individuals in the communications industry had somewhat parallel interests. They hoped to make new areas of the electromagnetic spectrum available for communication, or to develop means of aiding aircraft land in bad weather.

THE NAVAL RESEARCH LABORATORY

While the military organizations had been alerted to the potentialities that might be offered by research devoted to the study of reflections of short electromagnetic waves from moving objects as early as the 1920s, the pace they adopted was slow until the mid-1930s, in large part because of budgetary constraints. A special civilian group within the navy attempted to pursue the field as early as 1922 when reflections from moving objects such as motor vehicles were first detected. This alert group was located in the Naval Research Laboratory at Anacostia in the District of Columbia—a very special institution established in 1923 as the result of a proposal made by Thomas A. Edison during World War I. The initial staff was drawn from various naval laboratories and was placed under the directorship of A. Hoyt Taylor, who had been deeply involved in the advance of radio technology within the navy.

A unique feature of the laboratory was the high degree of freedom accorded the civilian members in selecting their research programs. Funding was, however, another matter. Taylor attempted to interest the navy in providing support for research and development in the meter range of wavelengths at the start, but he was unsuccessful. The civilian staff managed to carry on some of the research that was of greatest personal interest to them by working overtime on their own with whatever equipment they could find or assemble from stray parts. They were, for example, of considerable help to G. Breit and M. Tuve at the Carnegie Institution when the pair was engaged in probing the ionosphere with pulsed electromagnetic beams, as described in chapter 6.

The subject of airplane detection was revived at the laboratory in 1930 when reflections from moving aircraft were discovered accidentally during a series of experiments on direction finding. The scientists urged the naval command to provide support for further investigations, but this did not occur until 1933 when members of the Bell Telephone Laboratories, speaking at an open technical meeting, reported similar, independent observations. The importance of observations of this type for military purposes did not escape the attention of the British.

In any event, the U.S. Navy did decide in 1934 to allow Robert Page, a bright young member of the staff at the laboratory, to work at least part time on radar detection. The rest of the story is part of the technical history of the navy. Page not only became deeply immersed in the topic but eventually decided that pulsed methods of detection were preferable to those using continuous waves, which had been the focus of attention earlier. Overcoming many obstacles, by 1939 Page and his group had designed and put into operation a practical radar system using 385-megahertz (78 cm) radiation. In the meantime, industrial contractors were brought in to assist in further development and production.

THE SIGNAL CORPS

The Signal Corps Laboratory of the army at Fort Monmouth, New Jersey, had a parallel interest in the detection of moving objects. While maintaining links with the work at the Naval Research Laboratory, it initially focused its own research on the use of infrared radiation, with some degree of success. Eventually, however, it was strongly influenced by the success achieved by the navy with pulsed radio waves and established a strong radar laboratory of its own at Fort Hancock on Sandy Hook, close to Fort Monmouth. One of the young participants in the Signal Corps Laboratory was William D. Hershberger, who had been much concerned by the pace of German research in the microwave field. This was especially so when he came across Hans Hollmann's two-volume treatise and realized its implication for military research. He made certain that his superiors were well aware of the potential of the field and the need for more intense activity.

THE RADIATION LABORATORY

The pace of research on the development of radar in the United States accelerated abruptly as a result of steps taken at the highest levels of government following the fall of France in the spring of 1940. The Radiation Laboratory at the Massachusetts Institute of Technology was created soon thereafter (officially in November 1940) and was encouraged by a decision in Britain to share that nation's secret knowledge in the field of radar with the United States.[2] Lee A. DuBridge, on leave from the University of Rochester, was selected as the laboratory's director. Although the British were well ahead, the United States began to catch up very rapidly in all phases of the technology and soon became the major coordinator of most radar research for the Allies.

While some individuals in the Radiation Laboratory hoped for the development of a vacuum tube mixer that might replace the cat-whisker diodes, experience soon showed that the semiconductor devices could perform their task well, when they behaved. Unfortunately, the units obtained initially from sources in United States tended to differ radically from each other and to behave more erratically than those from Britain, which had their own irregularities. It was common in the early stages of the program for a radar operator to carry a number of diode units and search for the best, replacing it with another if and when the first stopped functioning or otherwise became unusable.

OUTSIDE HELP

In late 1940 or early 1941, while at the University of Pennsylvania, Frederick Seitz received a call from Lee DuBridge asking him to visit the Radiation Laboratory to discuss the state of available diodes.[3] Soon a group at Seitz's university, which initially involved Andrew W. Lawson, Park H. Miller, Robert J. Maurer, and a very capable graduate student, Marshall D. Earle, was busy at work under a contract arranged by the Radiation Laboratory. The group was joined later by Leonard Schiff, Simon Pasternak, and W. E. Stephens, and eventually by M. L. Lewis. They assembled test equipment and visited other groups such as the Bell Telephone Laboratories and the General Electric Laboratory to learn of related activities.

One of the first actions was the construction of means for producing special samples of aluminum-doped silicon, both to replicate the action of the diodes being used and to measure the electrical conductivity and the Hall effect as functions of temperature to characterize the semiconductor more completely. The added measurements of the Hall coefficient made it possible to determine the mobility and density of carriers. Lawson developed a set of miniature beryllium oxide crucibles with which to produce ingots. It soon became evident that silicon is an intrinsic semiconductor at sufficiently high temperatures with an activation energy of about 1.1 electron volts (eV) and that the aluminum additions produce an extrinsic hole conductivity with a much lower activation energy of the order of 0.1 eV or less (eventually found to be about 45 meV). The mobility of electrons at room temperature was determined by adding elements such as phosphorus, which rendered the materials n-type, and was found to be about 300 cm/sec per volt/cm; that of the holes was about 100 cm/sec per volt/cm.

Later work with germanium by a group at Purdue University, discussed below, gave an intrinsic activation energy of about 0.76 eV below room temperature and mobilities of about 2600 for electrons and 1700 for holes near room temperature in the same units.

BIDWELL'S MEASUREMENTS ON GERMANIUM

One of the earliest discoveries made by the group at the University of Pennsylvania was based on a literature search that followed Seitz's first visit to the Radiation Laboratory. The search was carried out to see if any careful measurements of the electrical properties of silicon and germanium had been made during the preceding years. It revealed that in 1922 C. C. Bidwell, then at Cornell University, had published an excellent set of measurements of the electrical conductivity and thermoelectric power of a purified specimen of germanium.[4] The specimen had been provided to him by L. M. Dennis of Cornell's Department of Chemistry. Unfortunately, the Hall coefficient was not measured at the same time so that it was not possible to separate in a precise way factors related to the density of carriers from those that influence their mobility. It was possible only to estimate the variation in carrier density with temperature. Bidwell's results, as reproduced from his paper, are shown on page 139. When the measurements of the electrical resistivity were replotted to provide a Boltzmann (Arrhenius) representation, in which the logarithm of the resistivity is displayed as a function of the reciprocal of the absolute temperature, a curve of the form shown on page 140 emerged. (This is actually a present-day, computer-based version of the Boltzmann plot.) The transformed plot showed clearly that germanium is both an extrinsic and an intrinsic semiconductor with a band gap of the order of 0.6 electron volts, in reasonably close agreement with the more precise value of 0.66 eV determined later and valid near room temperature. The analysis for germanium made it evident that silicon, a close relative, must have much the same properties.

The very steep drop in resistivity at temperatures above 200° C in both figures (at the left side in the figure on page 140) corresponds to the appearance of the intrinsic conductivity associated with the generation of electron-hole pairs as a result of thermally induced excitation of electrons from the filled to the empty band. The relatively gentle downward slope on the right-hand side of the latter figure, near minus 200° C, corresponds to the freeing of electrons or holes, or both, from residual impurities. In the absence of Hall-effect data, it was possible to state only that the trapping energy for such carriers is of the order of a hundredth of an electron volt.

The upswing of the curve in the range of 0° C, shown on page 140, can be attributed to the increase in scattering of carriers with rising temperature by thermally induced lattice vibrations. Such scattering is also responsible for the rise in resistivity above 650° C shown in Bidwell's original plot of his data (page 139).

THE BELL TELEPHONE LABORATORIES

During early visits to the Bell Laboratories, the team from Pennsylvania found a group working under J. H. Scaff, who was, along with his colleagues, much concerned about the eccentric behavior of the diodes (see chapter 13).[5] Scaff also recognized that a substantial part of the erratic behavior lay in the use of relatively crude, inhomogenous metallurgical-grade silicon, and he suspected that the main contaminant was phosphorus. To improve the situation, he had started attempts to purify the commercial silicon by fractional crystallization, hoping that he would obtain substantial segregation of the impurities. This process was greatly refined by W. G. Pfann in the postwar era in the procedure known as zone refining (see chapter 16).

THE MAJOR DUPONT CONTRIBUTION

Fortunately, the group at the University of Pennsylvania had had a more direct route toward purification—one that solved, for the time being, a major problem concerning the silicon diodes.

Soon after Seitz had joined the university in 1939, the Pigments Department of the DuPont Company near Wilmington, Delaware, had asked him to serve as consultant on a major problem it was facing. Prior to 1930, the standard white pigments used in paints were compounds of lead such as lead carbonate, which were known to cause a wasting disease, commonly known as painter's colic. It incapacitated and eventually killed house painters in their thirties and forties. Such pigments were being replaced by non-toxic titanium dioxide, preferably in the cubic, high-index of refraction form, rutile. At the time Seitz was called in, DuPont was using a wet batch process invented in Europe to produce rutile. There were two interrelated questions. Was there a better, simpler white pigment? If not, was there a simpler way of producing rutile of satisfactory quality? Seitz became deeply involved in this important commercial and health-related program backed by substantial resources. The imaginative research staff was prepared to try any reasonable approach aimed at solving the problem, whether chemical or physical.

By 1942, they had discovered that rutile pigment of excellent quality could be produced continuously by reacting titanium tetrachloride with oxygen in an appropriately designed furnace—a process still widely used. The only conceivable competitor to titanium dioxide appeared to be a strictly stoichiometric form of silicon carbide, a combination of the elements very difficult to produce in practical, commercial quantities because of the relatively high solubility of discoloring carbon in silicon carbide at high temperatures.

The problem of producing purer elementary silicon emerged in 1941, just as the research program at DuPont was nearing its peak, and it was enthusiastically taken on as a parallel program.[6] The enlarged group soon found that silicon that was pure to about one part in one hundred thousand with respect to significant impurities (so-called five-nines material) could be produced in powder and granular form at reasonable cost by reacting silicon tetrachloride with zinc in a vapor phase reaction. The level of carbon in the product may have been outside the estimated range of purity, but apparently it had no significant influence on the diode properties. The method used was proposed and tested by C. M. Olson.

The research group at the University of Pennsylvania received samples from the first batches of the DuPont material and promptly carried on a variety of experiments. Andrew Lawson tested the effect of various additional agents on the bulk electrical properties while the other members worked with him in producing test diodes and examining their behavior. They soon discovered that boron was an excellent additive agent for enhancing the extrinsic semiconductivity. Apparently the same discovery was made independently at the Bell Laboratories at about the same time.

J. W. Ryde of the British Thomson Houston Company attested to the relative virtues of the DuPont silicon in a statement made in 1941: "In July 1941, BTH tested BTL crystals against their own, and found that the American crystals were more uniform and superior in performance. They attribute this to the better quality of the DuPont silicon used by BTL. In 1944, BTH requested permission to import DuPont silicon for the production of high burnout X-band crystals such as CV253, it having been found that the yield when using British material was unacceptably low."[7]

GERMANIUM

Harper Q. North at the General Electric Laboratory was, during this period, hoping to resolve the purity problem through another route. He had decided that germanium might be a better material than silicon and developed interesting semiconductor devices with it. Silicon, however, became the preferred material for actual service in radar, once the purer form was available, since devices made from it were less temperature sensitive and more stable in other respects.

VACUUM TUBE ALTERNATIVE

It was, perhaps, very fortunate for the future course of silicon-based electronics that DuPont was so well positioned to take on the problem of producing a purer form of the element at such a critical moment. Had there been a substantial delay, the group at the Radiation Laboratory would probably have sought an alternative to silicon. Germanium might have been substituted, at least for a period of time. However, in view of the composition of the staff at the Radiation Laboratory, there probably would have been pressure to develop a vacuum tube diode for the centimeter range, pushing the technology to the limit. Moreover, such a program probably would have been reasonably successful, at least for the immediate wartime needs.

In this connection, Jack Morton of the Bell Telephone Laboratories did push vacuum tube development to the limit during the war, formulating and manufacturing a triode vacuum tube that was usable in the ten-centimeter range of wavelengths. A diode presumably would have been at least as feasible. It seems safe to say, however, that in view of Mervin Kelly's determination to see what could be done with semiconductors (see chapter 13), the program that led to the transistor would have been pursued unchanged after the war at the Bell Telephone Laboratories, initially with germanium. The interest in silicon would eventually have reemerged in view of the advantages it offered and which were well known to its own microwave group. The delay in timing of the invention of the transistor probably would have been small. (See also the French discover, page 174.)

CENTRALIZATION AND EXPANSION

By 1942, it had become clear to the leaders of the Radiation Laboratory that they were dependent upon semiconducting diodes as mixers for the foreseeable future. As a result, they expanded the research program as broadly as possible, including other groups from industrial, governmental, and academic laboratories. First in importance, they developed a strong guiding team at the Radiation Laboratory of the Massachusetts Institute of Technology itself under the leadership of Henry C. Torrey and Marvin Fox.[8] They were soon joined by Charles A. Whitmer, Hillard B. Huntington (who had earlier worked with the group at Pennsylvania), C. S. Pearsal, and Virginia Powell. The group did an excellent job of leading the entire program.

Andrew Lawson eventually left the University of Pennsylvania to join the Radiation Laboratory but became a member of a different part of the organization. During this period, Torrey focused much of his personal research on studies of the high-voltage breakdown or "burnout" of units and came to be dubbed "the crystal crackin' papa," inspired by a current popular song, "Pistol Packin' Mama."

The group in the Radiation Laboratory received much help, particularly in the early stages of the program, from leading theoretical scientists at the same laboratory, particularly Hans Bethe. They were also aided by some of the members of Edward Purcell's group on Fundamental Developments, such as R. H. Dicke, S. Roberts, and E. R. Beringer.

PURDUE UNIVERSITY

Karl Lark-Horovitz, head of the Department of Physics at Purdue University, soon learned of the program and developed a major effort under the auspices of the Radiation Laboratory. He and his group focused their attention on the properties of germanium and germanium diodes, using the Eagle Picher Company of Joplin, Missouri, as a source of supply of pure material. He visited the group at Pennsylvania and was provided with plans for the basic equipment being used there. His colleague Ralph Bray helped him guide the Purdue program.[9]

Although the physics department at Purdue had lost some of its faculty to other wartime programs, the residual staff, to which others had been added, was excellent and much interested in pursuing research on germanium in a team effort under good leadership. The group included S. Benzer, now a distinguished molecular biologist, Ralph Bray, Vivian A. Johnson, Robert Sachs, R. N. Smith, and H. J. Yearian. K. F. Hertzfeld of Catholic University of America in Washington, as well as Esther M. Conwell, a new graduate student at the University of Rochester, and V. F. Weisskopf, then also at the University of Rochester, were important consultants, adding much refinement to theoretical aspects of the research program.

The contributions of the research group at Purdue University were formidable in spite of the fact that the team focused its activities on germanium, which was mainly of secondary interest for the immediately applied work at the Radiation Laboratory. The research was thorough and directed as much toward fundamental studies of the characteristic properties of germanium as upon rectification. Lark-Horovitz obviously believed that silicon and germanium held great promise for the future of science and technology and wished to lay down as firm a foundation of fundamental understanding as possible under wartime circumstances, while at the same time serving the requirements of the radar workers. In any event, the work of the group was by no means without practical consequences. It developed, along with more basic studies, some remarkable germanium diodes that could withstand potentials of the order of one hundred volts in the reverse, low-conducting direction. Such diodes were useful in auxiliary radar circuitry and were never quite matched in performance for that purpose by silicon rectifiers during the war.

INDUSTRIAL CONTRIBUTORS

The industrial groups were no less creative. The Bell Telephone Laboratories continued research and development at all levels, as well as the production of diodes for in-house use. S. J. Angello, a former student at the University of Pennsylvania involved with copper oxide rectifiers at Westinghouse, soon shifted his interest there to silicon diodes. In addition, the Sylvania Electric Company agreed to become a major producer of units. Its liaison with the Radiation Laboratory and other members of the extended network was carried on by a remarkable young scientist-engineer, Nathaniel Rochester, who had started his career with the Radiation Laboratory working with crystal diodes and who left the team in 1943. J. R. Woodyard of the Sperry Company, who had been associated with W. W. Hansen at Stanford University, became involved in the program. His associates E. Ginzton and E. Sherwood also became participants in the activity.

THE REVIEW CONFERENCES

One of the notable features of this endeavor was a series of regular meetings of the participants in the program under the sponsorship of the Radiation Laboratory. At these meetings, research programs were reviewed and discussed freely. The meetings were held at various places, but usually at Columbia University, about every two months or so under secure conditions. They were by no means restricted to the leaders of the groups, so that they eventually involved a hundred or more individuals. Almost all the discussion was free and open, although the Bell Telephone Laboratories (see chapter 13), which had had a very large internal program with its own long-range interests in communications in mind, understandably limited its representation to matters most directly concerned with the consequence of its work for the Radiation Laboratory. Scaff was a steadfast contributor, usually joined by his colleague H. C. Theurer and occasionally by A. H. White.

It would be hard to overstate the great importance of these regular review meetings for the rational development of the entire program. Although the participants maintained friendly, open relations with individuals at the government laboratories, such as the Naval Research Laboratory, the Signal Corps Laboratory at Fort Monmouth, and the National Bureau of Standards, there was little if any direct contribution from them. Most scientists in those organizations were deeply involved in technical work closely related to battlefield operations.

Eventually, by 1944, Whitmer and Huntington were moved to other activities and the crystal program was merged, along with Torrey, into a group led by R. V. Pound.

POSTWAR WORK

The Purdue group continued its efforts after the war as a major activity of the department. Since a great deal of the research was ultimately cleared for publication, it occupied a prominent place in the open literature. Ernest Braun made good use of the publications in preparing "Selected Topics from the History of Semiconductor Physics and Its Applications," chapter 7 in the historical survey book *Out of the Crystal Maze.*[10]

Esther Conwell completed graduate work at the University of Chicago, followed by a very rewarding scientific career in research and research management in both industrial and academic settings. After a brief period with the Bell Telephone Laboratories, she joined the GTE Laboratory in Bayside,

New York, rising to the position of head of the physics department. This was followed by a senior research appointment at the Xerox Corporation Laboratory in Webster, New York, with an adjunct professorship at the University of Rochester. She was elected to the National Academy of Sciences in 1996.

One anecdote related by John Bardeen deserves mention here. In 1947, after Bardeen and Brattain had discovered both field-effect and bipolar transistor action at the Bell Telephone Laboratories using point-contact electrodes with silicon and germanium, and while confidentiality was being maintained during the filing of patents, they visited Purdue as guests of the physics department. Lark-Horovitz, still deeply involved in the study of germanium, said to them: "There must be some way in which we can make a triode from these semiconductors. Do you have any suggestions?"

The key individuals involved in the work on semiconducting diodes dispersed at the end of the war, with some reshuffling of affiliations. H. C. Torrey and C. A. Whitmer joined the faculty of Rutgers University. Whitmer eventually moved to the National Science Foundation. M. Fox became the leader of the accelerator design group at the Brookhaven National Laboratory. A. W. Lawson went first to the University of Chicago and then to the University of California at Riverside. P. H. Miller first entered the aerospace industry and then joined the U.S. International University of California West, whereas R. J. Maurer spent most of his career at the University of Illinois. L. I. Schiff became chairman of the Department of Physics at Stanford and S. Pasternak became an editor of the *Physical Review*. W. E. Stephens continued at the University of Pennsylvania. H. B. Huntington became a prominent member of the physics department of the Rensselaer Polytechnic Institute. J. H. Scaff, H. C. Theurer, and A. H. White remained at the Bell Laboratories. H. Q. North joined the growing California aerospace industry, and E. L. Ginzton became a leader in the management of the Varian Corporation. N. Rochester remained in industrial activity.

Seitz left the University of Pennsylvania at the end of 1942 to become head of the physics department at the Carnegie Institute of Technology. Leonard Schiff, who had been much involved in the diode program, took over and carried on until he was called to Los Alamos in the spring of 1945. Seitz, however, continued to participate in the regular review sessions usually held at Columbia University until other wartime activities, which began in 1943, made attendance impractical.

DuPont never became a major supplier of silicon to the transistor manufacturers in the long run. The material that it produced for diodes was not sufficiently pure to be acceptable as the technology advanced. In the meantime, DuPont decided not to divert research effort away from its support of other commercial activities, particularly the development of polymers. This was an entirely reasonable decision for the time since polymer research was then achieving miraculous results—results that have also had a pervasive effect on the well being of society and the convenience of living.

SUMMARY

There is little doubt that the collective work on the semiconductors made it possible to utilize semiconducting diodes in radar systems in a completely practical and systematic way—the main objective of the research. The chemists, physicists, and metallurgists involved were rightfully proud of the fact that relatively small dedicated groups at universities and industry were able to contribute to this endeavor rapidly and at a critical time. They owed much, of course, to the special events taking place in one of the branches of the DuPont Company precisely at that time.

There were participants who had followed the advances in solid-state chemistry and physics almost from the start of the era in which the field was opened up by the new developments in quantum mechanics. They were thrilled by the new revelations concerning the two semiconductors. These semiconductors stood prominently as links between the metals and the well-recognized insulators with large band gaps. Understanding their properties required just the type of theory that had been under development for more than a decade. The work and the discoveries going with it provided a very special sense of fulfillment as the revelations emerged.

A historically significant photograph involving the development of microwave technology in the United States. Gathered about one of the first klystrons developed at Stanford University in the mid- to late 1930s are, clockwise, the inventor William W. Hansen *(standing on right)*; John R. Woodyard, an advanced graduate student who later became involved in the silicon diode program; the Varian brothers, Russell and Sigurd; and David L. Webster, the head of the physics department. (Courtesy of the Department of Special Collections, Stanford University.)

A. Hoyt Taylor, the electronics expert who headed the Naval Research Laboratory when it was formed in 1923 as the result of a proposal made earlier by Thomas Edison. (Courtesy of Herbert Friedman and the Naval Research Laboratory.)

Robert Page, who accepted the responsibility for exploring the potentialities of radar detection at the Naval Research Laboratory in 1934. He and his team worked with dedicated zeal and had systems that used 385-megahertz radiation operating by 1939. (Courtesy of Herbert Friedman and the Naval Research Laboratory.)

Lee A. DuBridge, who directed the Radiation Laboratory at the Massachusetts Institute of Technology during World War II. He became the highly effective president of the California Institute of Technology after returning briefly to the University of Rochester, from which he had taken leave in the autumn of 1940. (Courtesy of the American Institute of Physics Emilio Segrè Visual Archives.)

Andrew W. Lawson *(center)*, in a group photograph taken at the Radiation Laboratory toward the end of World War II. On the left is J. B. Horner Kuper, who later became head of electronics at Brookhaven National Laboratory. George H. Vineyard is on the right. He eventually became director of physics at Brookhaven. The original photograph appeared in *Five Years at the Radiation Laboratory* (Cambridge, Mass.: Massachusetts Institute of Technology, 1946; reprinted by IEEE for the International Microwave Symposium, 1991).

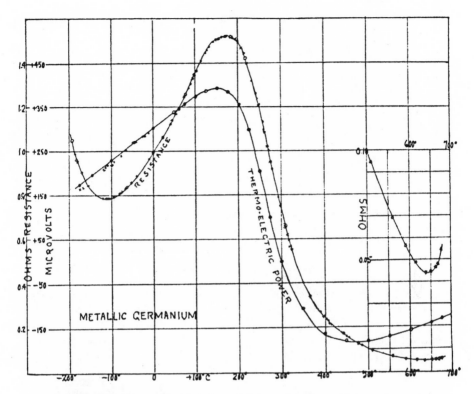

Measurements of the electrical resistivity and thermoelectric coefficient of a specimen of chemically pure germanium carried out over a wide range of temperature by C. C. Bidwell at Cornell University in 1922. The Hall coefficient was not measured. (From *Physical Review* 19 [1922]: 447.)

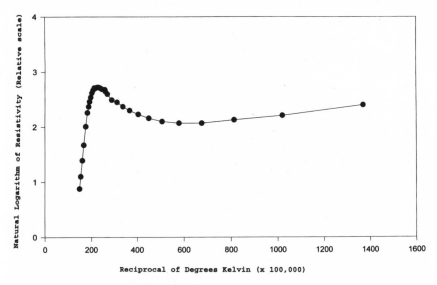

A present-day computerized Boltzmann plot of a portion of the data shown on page 139. The horizontal axis is the reciprocal of the absolute temperature multiplied by 100,000 for convenience, whereas the vertical axis is the natural logarithm of the resistivity, also in arbitrary units. The peak in resistivity for the specimen employed in the experiment occurs at about 170° C.

Jack H. Scaff, an experimental metallurgist who took on the problem of purifying silicon by means of fractional crystallization. (Property of AT&T Archives. Reprinted with permission of AT&T.)

C. Marcus Olson
(center), the DuPont
research chemist who
proposed the method
used to produce the
pure grade of silicon
required for the
crystal diode mixer
program. Olson is
accompanied by two
colleagues engaged in
chemical research,
Charles J. Carignan
(left) and George L.
Lewis *(right)*. (Cour-
tesy of C. M. Olson.)

Jack Morton of the Bell Telephone
Laboratories, who successfully
completed the difficult task of developing and manufacturing a triode vacuum tube which operated
at wavelengths of ten centimeters.
(Courtesy of the American Institute
of Physics Emilio Segrè Visual Archives, Physics Today
Collection.)

Henry C. Torrey, the member of the staff of the Radiation Laboratory who organized and coordinated the eventually widespread group of individuals and laboratories involved in the silicon and germanium diode programs, once it became clear that reliably effective units could be manufactured in adequate quantity. (Courtesy of H. C. Torrey.)

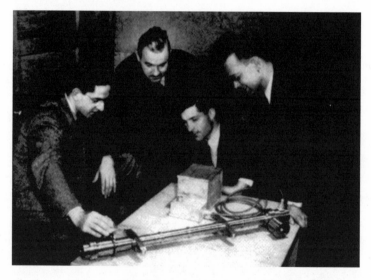

Charles A. Whitmer *(standing, center),* who worked closely with Torrey in managing the silicon diode program. Hillard B. Huntington, a member of the diode group, is seated. The individual on the far left is W. Selove; R. A. McConnell is on the far right. The photograph originally appeared in *Five Years at the Radiation Laboratory* (Cambridge, Mass.: Massachusetts Institute of Technology, 1946; reprinted by IEEE for the International Microwave Symposium, 1991).

CRYSTAL RECTIFIERS

By HENRY C. TORREY
ASSOCIATE PROFESSOR OF PHYSICS
RUTGERS UNIVERSITY

And CHARLES A. WHITMER
ASSOCIATE PROFESSOR OF PHYSICS
RUTGERS UNIVERSITY

EDITED BY

S. A. GOUDSMIT   LEON B. LINFORD

JAMES L. LAWSON   ALBERT M. STONE

OFFICE OF SCIENTIFIC RESEARCH AND DEVELOPMENT

NATIONAL DEFENSE RESEARCH COMMITTEE

FIRST EDITION

NEW YORK AND LONDON
MCGRAW-HILL BOOK COMPANY, INC.
1948

Title page of volume 15 of the Radiation Laboratory series of books summarizing wartime research on radar. This volume, prepared by H. C. Torrey and his colleague C. A. Whitmer, deals with the program on crystal diodes.

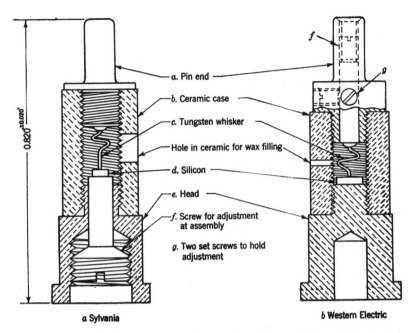

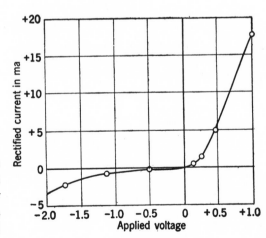

State-of-the-art silicon-tungsten diodes produced in 1942 by two of the partici-
pating companies. (From H. C. Torrey and C. A. Whitmer, *Crystal Rectifiers*.)

A typical current-voltage
curve for silicon-tungsten
diodes. (From H. C. Torrey
and C. A. Whitmer,
Crystal Rectifiers.)

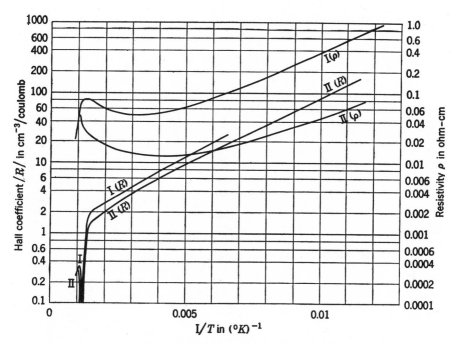

Resistivity (ρ) and Hall coefficient (R) for two specimens of silicon as a function of temperature. The specimens were p-type, containing additions of aluminum. These are Boltzmann plots similar to that shown on page 140 for germanium. The Hall coefficient is inversely proportional to the density of carriers. (From H. C. Torrey and C. A. Whitmer, *Crystal Rectifiers.*)

Karl Lark-Horovitz and a group of visitors attending a scientific meeting at Purdue University in 1942. From left to right are: *(front)* Wolfgang Pauli, Julian Schwinger, Edward Condon, Joseph Becker; *(rear)* Lark-Horovitz, William Hansen, Donald Kerst. (Courtesy of A. Tubis and Purdue University.)

A portion of the Purdue team that focused on the properties of germanium and germanium diodes. Front row *(left to right):* R. N. Smith, R. Bray, A. Ginsberg, W. W. Scanlon, P. B. Pickar; second row: R. M. Whaley, S. Benzer, V. A. Johnson, I. I. Boyarsky, H. J. Yearian; third row: E. S. Akely, A. W. McDonald, K. Lark-Horovitz, T. S. Renzema, I. Walerstein. (Courtesy of R. Bray and the *Lafayette Journal and Courier.* This photograph appeared in the November 3, 1945, issue of the newspaper.)

(Facing page, bottom) Ralph Bray, a major contributor to the program at Purdue University dealing with the exploration of the electronic properties of germanium and germanium diodes. This photograph dates from 1943 or 1944. Bray is at the bench testing the properties of a germanium diode. He continued forefront research at Purdue in this and related fields in subsequent decades. As the Purdue team's work advanced, the members focused part of their attention on the possibility of producing a triode device. (Courtesy of Ralph Bray.)

NOTES

1. See Henry E. Guerlac, *Radar in World War II*, 2 vols. (New York: American Institute of Physics, 1987), chap. 13, n. 10.

2. See 28-volume collection *The Radiation Laboratory Series*, ed. Louis N. Ridenour (New York: Massachusetts Institute of Technology and McGraw-Hill, 1948), for the history and activities of the laboratory.

3. See F. Seitz, *Physics Today* 48 (1995): 22; H. Ehrenreich, ibid., p. 28; F. Seitz, *On the Frontier: My Life in Science* (New York: American Institute of Physics Press, 1994).

4. C. C. Bidwell, *Physical Review* 19 (1922): 447.

5. See Sidney Millman, ed., *A History of Engineering and Science in the Bell System (1925–1980)* (Short Hills, N.J.: AT&T Bell Telephone Laboratories, 1983).

6. C. M. Olson, "The Pure Stuff," *American Heritage of Invention and Technology* (Spring–Summer 1988): 58.

7. Ryde quoted in E. B. Callick, *Metres to Microwaves* (London: Institution of Electrical Engineers, Peter Peregrinus, 1990), p. 95, n. 5.

8. See H. C. Torrey and C. A. Whitmer, *Crystal Rectifiers,* The Radiation Laboratory Series, vol. 15 (New York: McGraw-Hill, 1948).

9. See Karl Lark-Horovitz, "Preparation of Semiconductors and Development of Crystal Rectifiers," *NDRC Report,* div. 14, report 585 (Washington, D.C.: U.S. Office of Scientific Research and Development, 1946); R. Bray, K. Lark-Horovitz, and R. N. Smith, *Physical Review* 72 (1947): 530; R. Bray, ibid. 74 (1948): 1218; ibid. 76 (1949): 152; Torrey and Whitmer, *Crystal Rectifiers;* L. Hoddeson et al., eds., *Out of the Crystal Maze* (New York: Oxford University Press, 1992); P. W. Hendriksen, "Solid State Physics Research at Purdue," *OSIRIS,* 2d ser., 2 (1987): 237; R. Bray, "The Origin of Semiconductor Research at Purdue University," *Purdue Physics* 2, no. 2 (1990): 6; idem, "The Invention of the Point-Contact Transistor: A Case Study in Serendipity," *Interface* 6, no. 1 (1997): 24. The last-named paper merits special interest. The Purdue group was highly conscious of the desire to develop a triode. Bray discusses some of the major factors that gave the Bell Laboratories an advantage in the search.

10. See L. Hoddeson et al., *Out of the Crystal Maze;* see also Millman, *History of Engineering and Science.*

JAPANESE RADAR

$\mathcal{T}$HE JAPANESE RELATIONSHIP with microwaves and radar is a complex one. Although that nation's scientists and engineers became involved in well-designed, independent research in the fields, the work never came to truly significant focus. The indigenous talent and interest were high and were stimulated early, but they were not always well used.[1] The Japanese military, with its in-house and national industrial links, came to appreciate the vital role radar would play in the Pacific war, but this realization came much too late to achieve significant development and deployment. While the Japanese alliance with Germany provided access to early German pulsed equipment, this was essentially cut off by Allied intervention as the war proceeded. Eventually, nothing other than technical information, and not a great deal of that, was shared.

In a postwar paper, H. Kumagi, Koichi Shimoda, S. Lio, and J. Yuhara, who were carrying out microwave research at the Institute of Science of the University of Tokyo during the war, describe their wartime work on crystal diodes for use in instrumentation and as heterodyne mixers.[2] Their studies extended as far as the ten-centimeter range of wavelength. Within the family of eleven semiconductors they examined, silicon and pyrites proved to be most effective with appropriate whiskers—tungsten, in the case of silicon. Of the two, silicon possessed more stable behavior over a range of temperature extending to 60° C. The investigators were familiar with the work of H. Klumb and B. Koch published in 1939[3] and referred to it.

NOTES

1. See Ulrich Kern, "Die Enstehung des Radar Verfahrens: Zur Geschichte der Radar Technik bis 1945" (Thesis, University of Stuttgart, 1984), for further details on Japanese radar.

2. See H. Kumagi, Koichi Shimoda, S. Lio, and J. Yuhara, "The Crystal Detector Used for Microwave Applications" (in Japanese), *Physical Society of Japan* 2, no. 5 (1947): 176. We are indebted to Professor Shigeru Sassa for help in translating this paper.

3. H. Klumb, *Physikalische Zeitschrift* 40 (1939): 640; H. Klumb and B. Koch, *Die Naturwissenschaften* 27 (1939): 547.

THIRTEEN

THE BELL TELEPHONE LABORATORIES

$\mathcal{U}$NTIL THE BREAKUP of American Telephone and Telegraph under an agreement with Judge Harold H. Greene in 1984, the Bell Telephone Laboratories undoubtedly constituted the most productive research organization in the world.[1] The official designation of the laboratories dates from 1925 when two major research and development groups from AT&T and the Western Electric Company, the manufacturing unit, were consolidated in a building on the west side of New York with Frank P. Jewett as president.[2] The core of the research activity extended much farther back in history, with significant steps of institutionalization taken in 1907 and 1911. Jewett had joined the organization in 1904 and risen through the upper ranks. For example, he was a key research engineer in the development of the triode vacuum tube for telephone repeater stations.

As the largest and most advanced telephone system in the world, AT&T drew on a vast array of basic and applied science and engineering for its continuously evolving needs, as new, more effective, and more efficient approaches to communications developed. Through its own laboratories it has not only done much to promote those approaches but has shared the benefits with the world. As with any telephone system, its equipment is widely dispersed through a network. As a result of the extended outreach of that network, much of the equipment is exposed to a diversified range of climate and environment, from the heart of turbulent cities, with their labyrinthine underpinnings, to the depths of oceans. The system must span complex terrain by whatever technology represents the state of the art at a given time. In these days of satellite links, it must also deal with the harsh environment outside the atmosphere. Meanwhile, its equipment must be reliable and enduring.

The combined Bell Telephone Laboratories were designed to deal, both in-house and through contract arrangements when necessary, with all essential issues related to the basic scientific and engineering needs of the system, giving careful attention to economic factors along the way. In the course of the productive history of the laboratories, they have opened major new areas of science, such as the discovery of electron diffraction by C. J. Davisson and L. H. Germer in 1927, and have encouraged the development of many nascent areas of both science and technology opened elsewhere. They helped in the pioneering of numerous aspects of solid-state physics and chemistry that play an important role in the subject of this book, but this is only one element of their outreach. They have explored in profound ways every part of the electromagnetic spectrum of possible interest in the field of communications. In the course of this, A. A. Penzias and R. W. Wilson discovered, in 1965, the residue of the primordial gamma radiation produced in the early history of the creation of our universe—radiation that now lies in the microwave portion of the spectrum as a result of the continuing expansion of that universe.

Closer to everyday affairs, the leadership recognized early, in the 1930s, the great benefits that the telephone system might gain through the innovations offered by the budding field of polymer chemistry. These individuals selected William O. Baker, who later became the president of the laboratories, to take the lead in exploring this rapidly expanding field and in helping the organization make use of the extraordinary opportunities offered by it. Many major benefits have followed from this decision, including the development of inexpensive, durable, polymer-coated coaxial cable and, more recently, the exploitation of its fiber-optic counterpart, which is now on the way to spanning the globe.

PUBLIC INFORMATION

Although much of the research carried out at the laboratories was published in the open literature soon after it was completed and deemed to be of sufficiently general interest, some activities have been considered to be of such great importance to the future well-being of the company that they were classified as confidential. Naturally, details of such work were often not released until they were no longer considered to have special value, or were added to expand a common pool of knowledge that had been developed through other sources. In such instances, the release might also be made in order to give appropriate professional recognition to those in the organization who had been responsible for the early work in the laboratories. Not infrequently, professional historians would be given access to older files with the privilege of documenting and subsequently releasing knowledge of the details.

The importance of external recognition was probably valued less by those in industrial research laboratories than by academic colleagues, who rely upon public recognition for advancement and professional prestige. The physicist Edward U. Condon, who had worked at the Bell Laboratories for a period of time in the 1920s and was later director of research at the Westinghouse Research Laboratory in Pittsburgh and the National Bureau of Standards in Washington, D.C., once said: "The Bell Laboratories are like an island universe. They have so many interesting and absorbing activities going on at any given time that one could almost ignore the outside world, at least in a professional sense."[3]

THE TRIODE VACUUM TUBE

One excellent early example of the manner in which the Bell Laboratories guarded important work is demonstrated by the early history of the triode vacuum tube. Theodore H. Vail, who had been with AT&T previously and was brought back as president in 1907 to carry out a substantial reorganization, decided that telephone service should be extended throughout the country, and he called upon the technical staff to find ways of doing this—a task it achieved in time for the Panama Pacific World Exhibition in San Francisco in 1915.[4] In an autobiographical essay, Frank Jewett stated that he and the associated staff obtained relatively little sleep while the effort to achieve this goal was underway.[5] Effective "repeater" stations with suitable amplifiers of some kind were needed, and the audion, invented by Lee De Forest—and still a very primitive device—seemed to offer some promise if suitably developed. The patent rights for the tube were purchased from De Forest in 1912, and responsibility for managing such development was placed in the hands of H. D. Arnold.

In the meantime, Irving Langmuir and his colleagues at the General Electric Laboratory in Schenectady, New York, also carried out research and technical improvements on the triode independently.[6] Their work made it possible to understand and utilize its properties far more effectively than had been the case for the original version. When the General Electric Company attempted to patent its more highly developed version just before World War I, litigation developed concerning the status of the contributions made by Langmuir and his associates. First among the issues, it was clear that the telephone laboratories had done a great deal on their own, behind the scenes, to improve the properties of the tube. Second, Langmuir's quite basic original researches on the behavior of the electron space charge surrounding the heated cathode were judged to be fundamental in nature, akin to the discovery of natural laws, which the judge involved in the review considered not patentable. As a result, the General Electric Company was not awarded a priority patent. Fortunately, this decision did not inhibit widespread research and development in the field by a number of companies in the United States, Europe, and elsewhere, so that vacuum tube technology advanced rapidly, particularly during and after World War I. It was apparent that a large market would open. In 1919 several American companies, including the American Marconi Company, the Western Electric Company, the General Electric Company, and Westinghouse Electric Company, created the Radio Corporation of America (RCA) with governmental approval. It was designed to serve, among other things, as a patent pool for work on radio, presumably to quell extended legal controversy and enhance the rate of development nationally. The leadership role in the arrangement was provided by Owen D. Young, a counselor to AT&T.

SEMICONDUCTORS

Another area in which the Bell Laboratories took a deep, behind-the-scenes interest at a later time is related to semiconductors, long after semiconductors had ceased to be of great interest in wireless or radio communication—if we exclude the interest of youthful amateurs who could afford no more than a crystal rectifier set at start. Perhaps unfortunately, much of the history of this phase of work at Bell Labs was made available well after the primary developments leading to the transistor had occurred. Consequently, it is difficult to gain perspective regarding what were truly "first" discoveries for which the Bell Laboratories deserve credit since the field was being explored internationally at the same time—usually more openly than at the Bell Laboratories.

Apparently, the staff at the Bell Laboratories took special note of the invention of the large-area cuprous-oxide-metal rectifier by L. O. Grondahl and P. H. Geiger in 1926 (see diagram on page 161) and decided that semiconductors as a class of materials deserved special attention, which they were already attracting in Europe.[7] In the case of cuprous oxide, the rectifying effects were pronounced, the system was relatively rugged, and the electric current, unlike that emerging from the cathode of a vacuum tube, did not require a special source of heat. One of the prime movers in encouraging and sustaining this interest was Mervin J. Kelly, director of the vacuum tube department in the period between 1928 and 1934 and director of research of the Bell Laboratories starting in 1936.[8] He felt intuitively that semiconducting devices should have a significant future in the field of electronics and made certain that work in solid-state science was well funded and staffed as far as circumstances permitted. He believed that, at a minimum, semiconductors might become important auxiliary units and might possibly supplant some vacuum tubes. It was in this period that the Bell Laboratories hired William B. Shockley (1936), Dean Wooldridge, Gerald L. Pearson, Foster C. Nix, and others interested in the physics and chemistry of crystalline solids to join colleagues such as Joseph Becker, Walter Brattain, and Richard Bozorth, who were already engaged in related research.

MICROWAVE RESEARCH

During the 1930s, the Bell Laboratories became involved in the use of semiconductors through another pathway that proved to be of major significance, namely, microwave research—a natural step for an organization devoted to expanding the field of communications. In a volume published in 1962, George C. Southworth, then retired from the laboratories, describes the use of silicon as a detector, wavemeter, and standing-wave indicator in investigations involving coaxial conductors that were carried out in cooperation with Arnold Bowen and W. J. King.[9] They selected silicon as the choice material early in their work as a result of previous experience with it in the form of point-contact rectifiers during wireless days. Such studies were presumably made parallel to and independently of those of Hans Hollmann (see chapter 7). The team found at this stage that other well-known semiconductors such as copper oxide and galena (PbS) were relatively ineffective either because they did not rectify at the required frequencies or, as in the case of natural galena, they were too inhomogeneous.

Subsequently, in a desire to find what might ultimately be the best material to use, Southworth involved his colleague Russell S. Ohl in a very wide search that covered a hundred or so elements and compounds. Ohl finally concluded that, among those available, the silicon-tungsten combination was indeed superior, even though the best commercially available form, obtained from Eimer and Amend Chemical Supply Company and of German origin, was highly variable because of impurities.

In his memoir, Southworth mentions that a group of German engineers from Telefunken visited his laboratory in October 1937.[10] Among the names of the visitors, he includes that of a person whom he designates "Dr. Karl Rottgart." One wonders if this individual was Jürgen Rottgardt (see chapter 7), who had decided independently that the silicon-tungsten combination of rectifier is the best of those available to serve as a crystal detector in the microwave range. If so, the two groups might well have compared results openly at the meeting since neither would have been concerned or familiar with secret research dealing with radar mixers at that stage of the development of radar.[11]

In August 1939, just before the start of World War II, Ohl turned the problem of producing purer silicon over to Jack Scaff and Henry Theurer, two of the metallurgists in the laboratories.[12] They in turn decided to see what could be achieved by fractional purification from molten silicon. This method was partly successful, but it was not to become truly effective until extended for special uses in the postwar period by William G. Pfann in the procedure known as "zone refining." In this process, the specimen to be purified is passed through a succession of cycles in which melting and freezing occur so that it experiences a number of repetitive steps of fractional crystallization. For the method to be effective, the unwanted impurities must preferentially concentrate in the liquid phase.

THE P-N JUNCTION

Some of the relatively refined specimens of silicon obtained by Scaff and Theurer in their initial work were found to conduct by means of negatively charged carriers (electrons) and others by means of positively charged ones (holes), leading to the designations "n-type" and "p-type," terms used in the previous chapters of this book and that quickly gained currency at the time. Scaff and Theurer suggested that the carriers would be n-type if the predominant impurities were the pentavalent elements that appear in the fifth column of the periodic chart, and p-type if trivalent elements in the third column—a supposition supported independently by broader and more detailed studies during wartime research.

One of the ingots obtained by Ohl in this work actually contained a region that was n-type and another that was p-type, with a very sharp interface between the two—the so-called p-n junction shown on page 85. It was to play a crucial role in the ultimate development of the bipolar junction transistor (see chapter 14). The research group was astonished by the magnitude of the voltage that was generated between the two halves when the junction was illuminated, a striking example of the photovoltaic effect.

KELLY'S RESOLUTION

When Mervin Kelly was shown the junction found by Ohl and its remarkable photovoltaic properties, he decided, quite rightly, that the discovery was probably of momentous value to the electronics industry and that the knowledge should be retained at the highest level of company secrecy until its true significance could be explored, presumably when the Bell Laboratories were no longer so heavily overwhelmed with military work. That push for further development was inevitably left until 1945, by which time Walter Brattain and William Shockley, joined by John Bardeen, were active in the company and prepared to provide leadership. In a sense Kelly can be regarded as the spiritual father of the transistor, if not the actual inventor.

Until well into 1941, scientists who became involved in research on silicon and germanium diodes as heterodyne mixers for radar, as a result of requests from the Radiation Laboratory at the Massachusetts Institute of Technology, had relatively free access to the Bell Laboratories for discussions of work of mutual interest, particularly of work on cuprous oxide and the results of Scaff's initial experiments with fractional crystallization. Then an abrupt change in policy took place in order to protect the company's future interest in the field. Ralph Bown, speaking for Kelly, informed the group that henceforth the laboratories would confine their presentations to the formal meetings mentioned in chapter 11, gatherings that were sponsored by the Radiation Laboratory and usually held in relatively secure conference rooms elsewhere about the country. The links with the Bell Laboratories were greatly attenuated.

The doors opened more widely again in 1948 after patents were filed and the invention of the bipolar point-contact transistor by Bardeen and Brattain was announced. Only then was it possible to comprehend the intensity of interest and activity that had existed at the Bell Laboratories during the years since 1941. By that time, however, many of the most basic studies of the properties of silicon and germanium had been carried out relatively openly by other laboratories and become part of common knowledge. Only information on the highly innovative work on the transistor was truly novel. The historical studies of the basic research carried out at the laboratories during the more "closed" period are very valuable in showing the importance of that work and the perception of the leadership, but they do little to establish priorities. To paraphrase a comment made by Marconi on another occasion, the laboratories were clearly "there."

The "West Street" building of AT&T, constructed in the 1880s. It served as the central laboratory from 1925, when Frank Jewett was made head of a more consolidated research organization of the company, until the laboratories were moved to New Jersey after World War II. (Property of AT&T Archives. Reprinted with permission of AT&T.)

Frank P. Jewett, during his early period as director of the Bell Telephone Laboratories. He served as president of the National Academy of Sciences during World War II. (Property of AT&T Archives. Reprinted with permission of AT&T.)

Theodore H. Vail, in 1885 when he first headed the American Telephone and Telegraph Company. He left the post to seek other opportunities in Latin America and elsewhere when the board of the company decided to restrict telephone service to the vicinity of New York City. He returned to the post in 1907 when it became clear to the board that expansion and consolidation were necessary to prevent the development of chaos. (Courtesy of Katherine M. Hurd, Vail's granddaughter.)

The first official telephone call between New York and San Francisco, January 23, 1915. Theodore Vail, the head of AT&T and seated at the center, is joined by a group of staff members and various officials. John Carty, the chief engineer of the company, is also on the line, standing immediately to the right of Vail. Frank Jewett, who carried much of the burden of research on the development of repeaters, is standing in the right-hand side of the photo. He can be identified by the flower in his lapel. (Courtesy of Katherine M. Hurd.)

H. D. Arnold, who was placed in overall charge of the task of expanding the
telephone system nationwide. (Property of AT&T Archives. Reprinted with
permission of AT&T.)

Irving Langmuir, the brilliantly
creative physical chemist who spent
his career at the General Electric
Laboratory in Schenectady, New
York. Along with much other work
in chemistry and electronics, he
provided the first clear understanding of the nature of the electron
space charge surrounding a heated
filament. (Courtesy of the National
Academy of Sciences.)

Lars Grondahl, the inventor of the cuprous oxide rectifier. He was employed
by the Union Switch and Signal Company at the time. The photograph dates
from the mid-1920s, the time of the
invention. (Courtesy of the American
Institute of Physics Emilio Segrè Visual
Archives.)

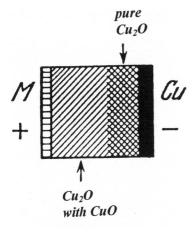

pure
Cu₂O

M   *Cu*

+   *−*

Cu₂O
with CuO

Schematic view of the large-area cuprous oxide rectifier. It is formed by controlled oxidation of metallic copper in such a way that there is a layer of cuprous oxide (Cu_2O) next to the copper, which serves as one electrode. The cuprous oxide is in turn accompanied by a layer of conducting cupric oxide (CuO) mixed with cuprous oxide upon which a second metal electrode is deposited. The rectifying junction lies at the interface between the cuprous oxide and the copper.

Mervin J. Kelly, director of research and then president of the Bell Telephone Laboratories between 1936 and 1959. While in charge of vacuum tube development in the late 1920s, he became very impressed with the merits of the cuprous oxide rectifier invented by Lars Grondahl and P. H Geiger. He can be regarded as the spiritual father of the transistor. (Courtesy of the National Academy of Sciences.)

G. C. Southworth, who pioneered microwave research in the AT&T system in the mid-1930s with a small group of colleagues. As a result of earlier experience, he began using point-contact silicon diodes as detectors and encouraged his close associate R. S. Ohl to explore the field to see what the best available combination of crystal and cat-whisker might be. (Property of AT&T Archives. Reprinted with permission of AT&T.)

R. S. Ohl, a radio and microwave pioneer who worked closely with G. C. Southworth. He searched independently for the best available crystal rectifier to use in the microwave region and decided that, among many combinations, silicon with a tungsten cat-whisker definitely formed the best pair. (Courtesy of Lillian Hoddeson.)

NOTES

1. For information on the laboratories, see Sidney Millman, ed., *A History of Engineering and Science in the Bell System (1925–1980)* (Short Hills, N.J.: AT&T Bell Telephone Laboratories, 1983).

2. A biography of F. P. Jewett appears in the *Biographical Memoirs of the National Academy of Sciences,* vol. 27 (1952), p. 239.

3. Comment to Frederick Seitz, circa 1932.

4. For a biography of Theodore Vail, see Albert Bigelow Paine, *In One Man's Life* (New York: Harper, 1921). See also the following papers by Lillian H. Hoddeson: "The Emergence of Basic Research in the Bell Telephone System, 1875–1915," *Technology and Culture* 22 (1981): 512; "The Roots of Solid State Research at the Bell Labs," *Physics Today* (1977): 23; "The Entry of the Quantum Theory of Solids into the Bell Telephone Laboratories, 1925–40," *Minerva* 18 (Autumn 1980): 422.

5. See Hoddeson, "Emergence of Basic Research."

6. A biography of I. Langmuir appears in the *Biographical Memoirs of the National Academy of Sciences,* vol. 45 (1974), p. 215.

7. See L. O. Grondahl, "A New Type of Contact Rectifier," *Physical Review* 27 (1926): 813; L. O. Grondahl and P. H. Geiger, "New Electronic Rectifier," *Transactions of the American Institute of Electrical Engineers* 46 (1927): 357.

8. A biography of M. J. Kelly appears in the *Biographical Memoirs of the National Academy of Sciences,* vol. 46 (1975), p. 191.

9. See G. C. Southworth, *Forty Years of Radio Research* (New York: Gordon and Breach, 1962); J. H. Scaff, "The Role of Metallurgy in the Technology of Semiconductors," *Metallurgical Transactions* 1 (1970): 562.

10. Southworth, *Forty Years,* p. 169.

11. Southworth's book also contains an extract from a letter he received in 1959 from Hans Mayer, a member of the staff of the Siemens Company who had visited Southworth's laboratory in 1938, just prior to World War II. In the letter, Mayer describes the mystification of the German electronics experts when they first examined the magnetron and related equipment found in a downed Allied bomber in 1943. As discussed earlier, in chapter 7, they had been prevented by government edict from developing any equipment in the centimeter range of wavelengths since it was regarded to be useless in pursuit of the war. Southworth, *Forty Years,* p. 174.

12. For discussion of Ohl, see Millman, *History of Engineering and Science;* Southworth, *Forty Years;* Scaff, "Role of Metallurgy." R. S. Ohl merits a prominent place in the history of the use of silicon in microwave research.

FOURTEEN

THE DISCRETE TRANSISTOR

$\mathscr{T}$HE STORY of the sequence of events leading to the invention of the transistor has been told many times, but it is important that the main features be highlighted here.[1] As mentioned in the previous chapter, when William Shockley returned to resume research at the Bell Laboratories in 1945, he was placed in charge of a research team that included the veterans Walter H. Brattain and Gerald L. Pearson, as well as John Bardeen,[2] who had just joined the Bell Laboratories, and who had spent most of the war years on research at a navy laboratory. The goal for the group, following Mervin Kelly's instructions, was to determine whether it was possible to develop a practical semiconductor triode. The extensive successful wartime research and development served to reinforce the vision of possibilities Kelly had held well before the war, a vision strengthened by the observation of the remarkable photovoltaic properties of the p-n junction produced accidentally in silicon by R. S. Ohl. Moreover, silicon and germanium, two highly versatile semiconductors that could conduct by both electrons and holes, were now available for the research.

FIELD-EFFECT DEVICE

In 1945, Shockley had attempted to develop what is today known as the field-effect transistor. A thin coating of semiconductor silicon was deposited by evaporation on an insulating backing. With a current of carriers flowing parallel to the sheet, a strong electric field was applied normal to the surface of the semiconductor with the aid of a metal electrode in the form of a plate close to and parallel to the plane of the semiconductor. It was hoped that the density of carriers in the semiconductor could be varied with the applied field as a result of changes in the charge induced in it. The experiment failed, as had indeed been the case for others who had tried a similar experiment previously and had failed to observe modulation.[3] Shockley still hoped to obtain a concept patent for this work. Unfortunately, this hope was blocked when similar concept patents obtained by J. E. Lilienfeld nearly twenty years earlier were uncovered.

Shockley, perhaps discouraged by the failure of this experiment, decided to turn a large part of his attention to research in other rapidly developing areas of solid-state physics, particularly in those associated with the study of lattice dislocations, and left most of the immediate continuation of research on the triode to Bardeen and Brattain and their associates. This was, in fact, one of Shockley's most creative periods of contribution to more general aspects of solid-state theory.

BARDEEN, BRATTAIN, AND SURFACE TRAPS

Bardeen, who had long contemplated the energy states associated with the surface of a solid, in part because his doctoral thesis at Princeton had been related to the subject, decided that Shockley's failure was probably related to the presence of a band of energy levels associated with the surface of the semiconductor that could serve as electron traps.[4] Levels of this kind could capture and release electrons in such a way as to produce a dipole field within the outer surface of the semiconductor that completely compensated for the variations in the modulating field Shockley had used. Moreover, Bardeen surmised that the time required for capture and release of electrons from the trapping states was sufficiently rapid at room temperature as to mask any true field effect in the range of modulating frequencies Shockley had used. This supposition was confirmed in two ways.

First, Shockley's version of the experiment was repeated by Pearson on an evaporated layer of silicon held at low temperatures, where the release time for the trapped electrons would presumably be much longer than at room temperature.[5] Some modulation of the current flowing in the silicon was actually observed, indicating that Bardeen's supposition was probably right.[6]

Concerned about Shockley's evaporated layer, Bardeen shifted to polycrystalline silicon and decided to focus next on possible changes that could be induced in naturally occurring depletion layers commonly situated at the surface of a semiconductor. Such a layer had been observed by Schottky and Spenke in earlier work,[7] as well as by those engaged in wartime research with diode mixers. The layer was of the order of one micron thick and was presumed to sustain some electrical conductivity by means of minority carriers, that is, electrons in p-type silicon. By way of illustration, one such depletion layer can be seen on the semiconductor side of the metal-semiconductor boundary shown at the bottom of page 82. In the latter case, it is produced as a result of the transfer of electrons from the semiconductor to the metal in the equilibrium configuration.

In various discussions Bardeen has pointed out that Walter Schottky, for whom he had great admiration, might have invented a version of the transistor in the late 1930s had it occurred to him that the presence of the surface depletion layer provided him with an opportunity to introduce a third electrode that could, for example, influence the concentration of minority carriers flowing in the boundary layer.

Bardeen was prepared to assume initially, as a tentative working hypothesis, or model, that any observed changes in injected current were to be associated with the effect that variable applied fields might have on the flow of carriers within the depletion zone, a hypothesis to be confirmed or altered with the advance of the research program.

Using a point-contact electrode, Bardeen and Brattain injected holes into the thin depletion layer at the surface of the semiconductor. From there the carriers migrated to a large-area electrode, termed the base, on the opposite surface after spreading laterally near the initial surface.

One configuration used is shown schematically on page 179. The specimen is mounted on a large-area metal contact. Two electrodes are placed on the top surface. One of these is a well-insulated point electrode that makes contact with the silicon at its tip. The second is immersed in an encapsulated, conducting, electrolytic solution that also makes contact with the silicon. The purpose of the solution is to provide both a conducting link between the semiconductor and the electrode immersed in it and to furnish ions that it was hoped would in some way tie up and neutralize the influence of the surface trapping states, as fortunately proved to be the case. The suggestion of using the electrolytic solution was made to Brattain by one of the physical chemists in the laboratory, Robert B. Gibney. A laterally spreading hole current was made to flow between the point-contact electrode and the base electrode, as shown in the figure. The depletion layer is also illustrated. A field effect was observed at room temperature with this system when the potential of the electrolytic electrode was varied relative to that of the base.

The evaporated layer of polycrystalline silicon used in the experiment possessed a high degree of lattice disorder, which lowers the electrical conductivity. As a result, Bardeen and Brattain decided to shift their studies to a single crystal of n-type germanium, which was made available to them, and to carry on measurements with modifications of the system. They started with similar use of the electrolytic solution. The specimen of n-type germanium also possessed a thin inversion layer, of opposite sign, at the surface. The field effect was observed more prominently in single-crystal germanium than in polycrystalline silicon.

In other words, Bardeen and Brattain had demonstrated that the concept of the field-effect transistor was sound. To achieve a practical triode using it, however, one needed to eliminate or control the surface trapping states. This was eventually accomplished by others, initially at the Bell Telephone Laboratories,[8] using silicon. In practice the surface is given an appropriate acid etch that removes contaminating foreign atoms that may provide trapping sites. Then a stable layer of silicon dioxide is deposited on the treated surface to serve as a passivating shield. In fact, by the 1970s the so-called metal-oxide-semiconductor field-effect transistor (MOSFET), which had undergone much further development over the years, had become a major workhorse of the trade. It offered many advantages, including relative simplicity of manufacture and more efficient use of power. Today, it dominates the field for the vast majority of semiconductor applications, in both memory and logic chips.

The elucidation of the importance of such surface states, which had plagued attempts in earlier, more primitive, times to develop a field-effect semiconducting triode, or to develop a reasonably precise theory of the rectifying blocking layer, must indeed rank as another of Bardeen's monumental contributions to the advance of solid-state electronics.

THE INVENTION OF THE POINT-CONTACT BIPOLAR TRANSISTOR

The use of the electrolyte shown on page 179 greatly limited the range of frequencies that could be used to modulate the current. As a result, Bardeen and Brattain decided to replace it with a circular disk of gold foil containing a central hole intended to accommodate the insulated point-contact electrode that would provide the current to the base of the primitive germanium device structure. The disk was mounted on a layer of germanium dioxide intended to insulate the gold disk electrically from the semiconductor. It was hoped that the electric field applied to the disk could be made sufficiently strong that it would overcome the influence of the surface traps. The insulation, however, proved to be ineffective; the gold foil actually made good electrical contact with the semiconductor.

As the investigators began examining the characteristic features of the new, presumably failed, arrangement, however, they made a remarkable discovery. When the point-contact electrode that was initially intended to be the source of the spreading current was moved sufficiently close to the gold electrode, and the potential of the point electrode was adjusted appropriately relative to the gold electrode, being more positive for the given situation, a current of presumably minority charges (in this case holes in the n-type germanium) could be made to flow from the point-contact electrode to the edge of the gold foil. Moreover, it was then found that the current of majority carriers flowing between the collecting electrode and the base could not only be substantially greater than the influx of minority carriers from the point-contact injecting electrode, but could be modulated by varying the potential of the injecting electrode and thereby modulating the flow of minority carriers. An entirely new concept had entered the picture, namely, the amplification and modulation of the flow of majority carriers from one electrode by the flow to it of minority carriers that have been injected into the semiconductor from another electrode. Bardeen and Brattain evidently had discovered far more than they had bargained for in starting the experiment, and had much to explore and learn.

The full importance of this partly chance discovery was obviously not lost on the pair. They immediately abandoned the gold leaf and employed two closely spaced electrodes separated by a distance of the order of fifty microns. One electrode was used to inject holes into the crystal; the other served to collect them. They were designated the "emitter" and the "collector." The current flowing between them was modulated by varying the potential of the injecting electrode relative to the collector and the base. The collector electrode could also serve as a source of majority carriers, electrons in the given case, to the base.

A schematic representation of a more highly developed demonstration model is shown on page 180. The modulated flow of minority carriers from the emitter to the collector, achieved by varying the potential between the two point-contact electrodes, modulated, in turn, the flow of majority carriers from the collector to the base, and also provided current amplification. A triode that employed an entirely new principle was born. It eventually came to be called the bipolar point-contact transistor. It was to have far-reaching consequences since it clearly opened up an entirely new area of exploration in the search for practical triodes based on the use of semiconductors.

It should be noted that the electrode and potential arrangements on the collector correspond to the reverse biased, or "blocking," mode of the metal-semiconductor junction, viewed as a rectifier. However, the space-charge field associated with the arriving minority carriers lowers the blocking barrier and permits the flow of majority carriers to the base and the positively charged emitter.

Bardeen and Brattain were amply prepared to exploit the discovery and were soon achieving voltage gains of about one hundred and current gains of about forty in suitably designed circuits. A carefully prepared, functioning experimental version, based on minority carrier injection, was demonstrated to several managers, led by Ralph Bown, on December 23, 1947, just a week after the momentous new discovery. Patent applications were filed.

The name "transistor" for the device was proposed by John R. Pierce of the Bell Laboratories in May 1948, following Brattain's request that he give thought to the matter. The name is based in part on the fact that the transfer of charge from the emitter to the collector in this carrier-injection, current-driven device involves the presence of a transfer resistor between input and output. Thus it is a "trans-sistor."

The official announcement of the discovery of the point-contact bipolar transistor took place in New York City during the summer of 1948. The units demonstrated at that time could operate in the range up to fifteen megahertz. (Frederick Seitz was giving lectures at Columbia University at the time and had the privilege of attending the ceremony.)

It should be added that nature was in its way very kind to the investigators in the sense that the recombination of electrons and holes involves indirect transitions (see chapter 5) in both silicon and germanium. That is, lattice vibrations play an essential role in permitting recombination. Bardeen and Brattain benefited from the advantage that the minority carriers in their systems had particularly favorable long lifetimes.

Had the discovery that a majority carrier current can be modulated by the injection of minority carriers not been made, it seems likely that the investigation of field-effect devices would have continued unabated and eventually proven to be very successful, as was ultimately the case, since the basic ground for exploiting them had been broken.

COMPLEXITY

The bipolar transistor actually is a complex, three-electrode, three-voltage device in which the operation depends upon the choice of potentials applied to the emitter, the collector, and the base. It can, and usually does, serve as a current multiplier because of the ability of the carriers of opposite sign to screen one another. Many details of its operation are given in Shockley's 1950 book, which was written after more thorough understanding had developed. To quote Shockley (in a passage he italicizes):

> In a semiconductor containing substantially only one type of carrier, it is impossible to increase the total carrier concentration by injecting carriers of the same type; however, such increases can be induced by injecting the opposite type since the space charge of the latter can be neutralized by an increased concentration of the type normally present.
>
> Thus we conclude that the existence of two processes of electron conduction in semiconductors, corresponding respectively to positive and negative mobile charges, is a major feature in several forms of transistor action.[9]

The operation of the field-effect transistor does not violate these principles in a commonly used form of the device. In that form, the applied modulating field alters the volume within which a fixed density of majority carriers may flow by altering the boundaries of inversion layers (see, for example, page 181).

SHIVE'S EXPERIMENT

Since the new invention had been achieved in a somewhat serendipitous manner, many questions concerning its inner workings were immediately raised. For example, Bardeen wondered if the charge associated with the injected minority carriers was by chance compensated by image charges in surface traps. He also wondered if the holes actually were moving only through the thin depletion layer at the surface or also flowed through the bulk n-type semiconductor. Shockley, who had made a theoretical study of the possible flow of charges in a p-n junction (see chapter 5) soon after returning to the laboratory at the end of the war, suggested that it was possible that a substantial fraction of the holes were moving through n-type material and that space-charge shielding of the majority carriers by the minority carriers was responsible for the variation in majority current at the collector associated with the variation in flow of minority carriers.[10]

As mentioned earlier, proposals that the holes were migrating through the bulk n-type volume of the semiconductor required that the lifetime for recombination of electrons and holes be sufficiently long under the conditions involved in the experiment that the minority carriers could survive the transit from one point electrode to the other. That this was, very fortunately, the case was demonstrated two months later by an experiment carried out by John Shive using a very thin layer of n-type germanium.[11] He found not only that a hole current injected on one side of the layer could be carried across to an electrode on the opposite side, but that the current could be modulated by varying the potential on the injector and a third electrode, in accordance with Bardeen and Brattain's discovery. All expressed relief that the puzzle was solved.[12] It was at this point that Shockley proposed a new version of the bipolar transistor, namely, the bipolar junction transistor, to be discussed in a subsequent section.

THE BARDEEN-BRATTAIN CHOICE

In retrospect, it is clear that the pathway followed by Brattain and Bardeen in using point-contact electrodes was a highly judicious one. It involved the simplest of equipment, fairly direct physical concepts, and a minimum of complex chemistry and metallurgy. While it is abundantly evident that Bardeen was the right individual at the right time for initiating a successful program at the Bell Laboratories, it must also be emphasized that the subsequent follow-through by the other dozen or so members of the staff involved in the research was both imaginative and thorough. The ultimate successful development of the transistor must be regarded as a reflection of the excellent teamwork that characterized the activities of the Bell Telephone Laboratories at its peak, each member providing imaginative input to obtain greater understanding and advance the development of the transistor. In fact, the laboratories probably were the main source of technical innovation in the field for the next decade.

THE BIPOLAR JUNCTION TRANSISTOR

Once the experiments carried out by Bardeen and Brattain and their colleagues were successful, and it was understood that a minority carrier could, under favorable circumstances, survive a transit across a region in which carriers of opposite type were dominant, Shockley, as mentioned, refocused his attention on further developments based on the underlying concepts involved in the point-contact transistor. He soon proposed an alternate version of the bipolar transistor, based on use of the p-n junction. In it, one p-n junction would serve as emitter, that is, the source of injected carriers, and another oppositely arranged p-n junction would serve as collector. This version of the transistor, termed the bipolar junction transistor, led the way to the next, more general, phase of development of semiconductor electronics, namely, the production of discrete transistors that were treated as replacements for, or complements to, vacuum tubes.

The most fundamental of Shockley's designs involves a central region of n-type or p-type material, to which the base electrode is attached, and which is sandwiched between two regions of opposite type that have their own electrodes (see lower diagram, page 180). Thus the composite unit contains two p-n junctions bordering each side of the center, or base, region. If an appropriate bias voltage, relative to that on the base electrode, is applied to one of the end regions—the emitter—what are majority carriers in it can be injected into the base region, where they become minority carriers, as in the point-contact bipolar transistor (see same page). Under ideal conditions, these charges can be made to flow to the other junction and on into the collector either by diffusion alone or additionally influenced by an applied field. The actual number of carriers transmitted from the emitter to the collector will, among other things, depend upon the difference in voltage between the emitter and collector, and on the potential between the emitter and base. Thus the minority-carrier induced current can be modulated by varying the three voltages, somewhat as in the bipolar point-contact transistor. The results obtained will depend upon the way in which the potentials are arranged and modulated. In the initial versions of the device, the migration from emitter to collector depended upon diffusion.

In one very simple arrangement that can lead to power amplification, the potential between the base and the emitter is modulated under conditions in which the emitter is biased so as to emit what become minority carriers into the base. Under ideal conditions, these carriers flow to the collector, which has a relatively high reverse bias potential that restrains the flow of the majority carriers from the base to the collector but does permit the minority carrier current to flow readily into the collector. The power gain of the system is approximately equal to the ratio of the bias potentials of the collector and emitter, although allowance must be made for the loss of minority carriers in transit, as well as other types of losses.

Other, comparably direct, arrangements of electrode potentials involving feedback to the emitter make it possible to obtain significant current amplification either alone or in combination with power amplification.[13] The potential of the new proposal was enormous. It opened the doorway to the truly major development of the discrete transistor and, ultimately, the integrated circuit.

Shockley's proposal, which employs large contact areas and volumes and permits the use of larger minority currents, proved to be far more flexibly useful than the point-contact transistor in major applications once the technology for producing it had been developed. Nonetheless, the principles upon which its design was based probably would not have occurred to the members of the team had the basic discovery of the point-contact version not happened as a consequence of the careful research by Bardeen and Brattain, described above. Moreover, even if the concept of minority carrier injection and modulation had, by chance, been proposed independently, the development of a demonstrable transistor would not have been achieved nearly as directly.

HAZARDS TO MINORITY CARRIERS

The minority carriers that migrate across the base region carry the primary transport current of the device. Under unfavorable circumstances, the carriers could be annihilated by combining with those of opposite sign, which, being in the majority, are relatively abundant in the region. Or they could become trapped at foreign atoms or other imperfections. Such trapping would not only impede the flow of current, as well as enhance the probability of annihilation, but would interfere with the rate at which the minority current could be modulated. It follows, as mentioned earlier, that high standards must be placed upon the purity and degree of perfection of the crystals of silicon or germanium used in fabricating such units. Two important types of lattice imperfection are shown on page 181. The imperfection on the left, termed a screw dislocation, has a central axis about which the orthogonal crystal planes form an extended spiral staircase. The lattice is distorted along the axis and can provide harmful trapping sites for minority carriers. The imperfection on the right has the effect of disorienting two sections of the crystal relative to one another, much as is the case at the boundary between two crystals of the same material (a so-called grain boundary). Such an imperfection contains many lines of distorted crystal that can interfere with the operation of a transistor.

THE FIELD-EFFECT TRANSISTOR AND THE BARDEEN PATENT

Early in 1948, after carrying through some of the seminal research on the point-contact bipolar transistor described earlier in this chapter, Bardeen conceived of special forms of field-effect transistor that were suggested by the initial work on the field effect carried out with Brattain. He applied successfully for what can be regarded as a broad basic patent on the device on February 26, 1948. The patent was granted on October 3, 1950.[14] In the several versions, the variable applied field could be used to vary the flow of current in duplex semiconductor systems containing normal and inversion layers. The history of his invention was related briefly by Bardeen in an interview carried out by Japanese television.[15]

INDUCED DEPLETION AND THE JUNCTION FIELD-EFFECT TRANSISTOR

If, as shown schematically on page 181, the voltages on electrodes on appropriately oriented p-n junctions on either side of a layer of semiconductor are altered relative to that of the base, minority carriers can flood into the base region and, under appropriate circumstances, replace some of the majority carriers. The corresponding effect in a bipolar junction transistor is illustrated on page 182.

The lower figure on page 181 actually illustrates a modified version of the field-effect transistor that is based more directly on the use of p-n junctions and the motion of the boundaries of inversion layers than is the case for the devices described in the Bardeen patent.[16] The first version was made by G. C. Dacey and I. M. Ross in 1953, following a suggestion by Shockley. In the case illustrated, the gate electrodes are in direct electrical contact with the p-n junctions so that the behavior of the inversion region is determined by the flow of current from the electrodes into the junctions. Alternatively, a layer of silicon dioxide can be used to insulate the electrodes from the semiconductor system, in which case the behavior of the inversion layers is determined by the voltage on the electrodes. This is the basis for an early useful metal-oxide-semiconductor field-effect transistor (MOSFET), termed the junction field-effect transistor. With the development of fabrication technology, it has been replaced by the so-called complementary field-effect transistor (CMOS) presumably covered by Bardeen's early patent.

Inversion or depletion regions can find other important uses, as in charge-coupled devices (CCDs) (see chapter 17).

FRENCH TRIODE INVENTION

In 1946 a French-organized team, stationed at a laboratory near Paris under the auspices of Westinghouse F. and S., a French company, began research aimed at developing germanium diodes for microwave use. Herbert Mataré, who had worked with Telefunken in Germany during the war, was invited to join the French team. He brought with him much experience and information as well as relatively pure but somewhat polycrystalline germanium specimens provided by H. Welker. Early in 1946, the group, stimulated by Mataré, helped extend the experimental program in an attempt to develop a triode using two point-contact probes along with a base electrode. They were successful in achieving an effect in some instances when one of the probes was placed at an intercrystalline boundary, presumably matching conditions of minority carrier injection similar to those achieved by Bardeen and Brattain in the bipolar point contact transistor. Minister Thomas of the French post office, which was supporting part of the work, expressed great interest in the new developments. A patent describing the triode effect was prepared and submitted in August 1948 and awarded in 1954. The discovery received much attention in the French technical press of the period. It clearly would have stimulated extended worldwide interest and activity had the publicity associated with the contemporary successful research program at Bell Telephone Laboratories not been overwhelming. We are indebted to Pierre Aigrain as well as Herbert Mataré for discussions of the French work. (See also page 103, note 8.)

SILICON VERSUS GERMANIUM

Initially it appeared that silicon and germanium might end up as equal partners in future practical developments. Further research, however, placed silicon far in the lead for several good reasons. First, it is plentiful in nature; supply was not a problem. Second, silicon devices can operate over a wider temperature range than can germanium devices. Third, silicon can be produced in very pure form if one takes the precaution to use very pure reagents in its preparation. Fourth, Gordon K. Teal eventually demonstrated that excellent single crystals could be produced in commercial quantities.[17] Finally, silicon oxide proved to be a fine, stable electrical insulator that could also serve as a good chemical barrier.

LICENSING AND KELLY'S DECISION

In keeping with his farsightedness, Mervin Kelly decided that further development of the transistor should not be confined to the Bell Laboratories alone, but should, for the sake of broad, even revolutionary, progress be carried on as widely as was reasonably possible.[18] This viewpoint was opposed initially by the Department of Defense, which had held that the new technology could be so important for national security that its existence and further development should be maintained as a major national secret. Ralph Bown took on the difficult task of convincing the department that its viewpoint would be counterproductive, and eventually he succeeded. Thus, licensing proceeded in 1951. The licensing fee for use of the basic patents was set at the modest figure of $25,000 so that access to the technology would be placed in the hands of an interested but carefully selected set of institutions that could be expected to provide new ideas and techniques for the advance of the technology. The opportunity was picked up with alacrity, particularly by corporations that were looking for diversification or by those that felt endangered by a new development. The end results were truly dramatic.

Among the first significant products, based on the use of germanium transistors, were light, low-powered hearing aids, which were an immediate success.

TEXAS INSTRUMENTS, IBM, AND OTHER PARTICIPANTS

Texas Instruments Incorporated, based in Dallas, Texas, decided in 1951 to make the transistor part of the central core of its business.[19] Prior to World War II, the company's main activity was contract geophysical prospecting for oil using seismic technology, a technology involving procedures that the company subsequently applied to the search for submarines during the war. AT&T was at first reluctant to give a license to a small company with this limited background but was finally persuaded to do so. It soon turned out that the decision was fully justified as a result of the leadership provided by a remarkably capable and perceptive engineer-businessman, Patrick E. Haggerty, who had joined the company after the war ended. Texas Instruments soon became one of the most highly innovative organizations struggling with the problems of design, fabrication, and use of the new devices.

International Business Machines (IBM), at the other end of the spectrum, a company that had a deep investment in the production of office equipment, decided that it could not ignore the field.[20] Its senior management took heed of the advice of John von Neumann, then at the Institute for Advanced Study in Princeton, regarding the coming revolution in computing through the use of programming appropriate to digital electronic circuitry. Wisdom required that the company needed to have a strong position in the field. In the course of events IBM developed several outstanding research centers that rivaled the Bell Laboratories in specialized areas.

Instrument and equipment makers such as Hewlett-Packard and Motorola realized that they could not afford to ignore such revolutionary new developments and also became major participants.

DUPONT SILICON AND ALTERNATIVES

Soon after the invention of the transistor, the Bell Laboratories asked the DuPont Company to resume production of silicon of the grade that had been so useful for diode mixers during wartime. The manufacturing plant, which was built at Brevard, North Carolina, operated for only a very short time in the 1950s because a rival process developed by the Siemens Company, and based on the reduction of pure trichlorosilane with hydrogen, produced a material of higher quality that became the staple product. DuPont did not attempt to offer a product of comparable quality and regain market share. The decision was quite reasonable at the time. The company was then deeply involved in the development and production of chemical polymers as commodities and had made that field its major focus of attention. Moreover, it was not obvious at that time that the production of slices of perfect single-crystal silicon would eventually grow into a multi-billion-dollar business.

William B. Shockley, who was selected to lead the research triumvirate that was assigned the ambitious task of developing a triode based on semiconductors. (Courtesy of the National Academy of Sciences.)

John Bardeen, who joined the Bell Laboratories at the end of World War II and came with much experience in both physics and electrical engineering. He and Walter Brattain discovered the importance of minority carrier injection and invented the bipolar point-contact transistor based on it. This invention led the way into the transistor age. (Bardeen Family photograph.)

Walter H. Brattain, a veteran of the Bell Laboratories who had much experience with solid-state physics and chemistry, particularly with cuprous oxide rectifiers. (Property of AT&T Archives. Reprinted with permission of AT&T.)

J. E. Lilienfeld, who obtained several concept patents on a field-effect transistor nearly twenty years before the work on the transistor started at the Bell Telephone Laboratories. The patents created interference with Shockley's application. The photograph is taken from Lilienfeld's U.S. naturalization documents. (Reprinted with permission from *Physics Today* [May 1988]: 87. © 1988 American Institute of Physics.)

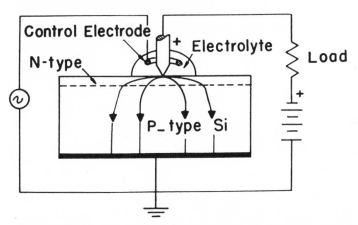

Schematic illustration of one of the experimental arrangements employed by Bardeen and Brattain in a room-temperature test of Bardeen's proposal that the surface-trapping states for electrons (or holes) were responsible for the difficulties encountered in developing a field-effect transistor. The current from the point-contact electrode to the base could be modulated by varying the potential of the electrolytic fluid relative to that on the base. This diagram, incidentally, also displays the type of inversion (or depletion) layer at the upper surface of the semiconductor that the investigators used in the experiment with polycrystalline silicon. (This figure was used by Bardeen in lectures describing his early work. Courtesy of N. Holonyak.)

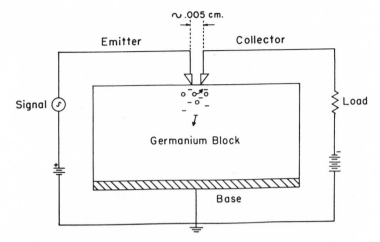

Schematic representation of the final, demonstrated, version of the bipolar point-contact transistor with the emitter and collector separated by about fifty microns. The germanium is n-type and the emitter, which is forward biased, injects holes, the minority carrier. The associated experiments demonstrated the importance of minority carrier injection. (Courtesy of the Bardeen Archive of the University of Illinois and Nick Holonyak.)

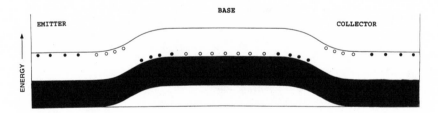

Schematic view of Shockley's bipolar junction transistor. It consists of two p-n junctions, in mirror-image conformation. In the case shown here, a p-type extrinsic semiconductor is sandwiched between two n-type semiconductors. If an appropriate negative biasing potential is applied to the emitter relative to the base—central—portion, electrons will flow into the upper empty conduction band of the base region where they are minority carriers. The magnitude of the minority current will be dependent upon the potential difference between the emitter and base, as well as upon the potential gradient between the former and the collector. Alternate arrangements of electrodes can be employed to achieve power amplification, current amplification, or both.

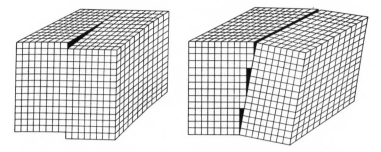

Two types of crystal imperfection. Apart from chemical impurities, defects in the lattice structure can interfere with the flow of minority carriers in the base region of a transistor. (Courtesy of Robert G. Hibberd. Reprinted by permission of Texas Instruments Incorporated.)

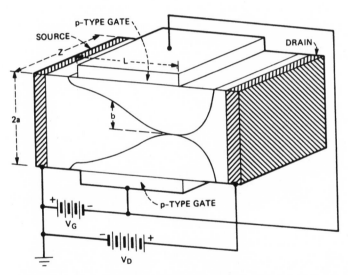

A schematic diagram showing the essential features of one form of field-effect transistor as developed by Ian M. Ross and G. C. Dacey at the Bell Telephone Laboratories in 1954, as a result of a suggestion by Shockley. (Property of AT&T Archives. Reprinted with permission of AT&T.)

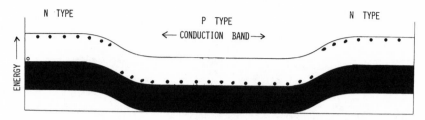

The depletion effect as illustrated in a bipolar junction transistor. If the voltage on the base of the type of transistor shown on page 180 is made strongly positive relative to the emitter, as shown in this diagram, electrons will flow into the base region from the emitter and collector depleting the base of some of its majority carriers, in this case holes.

NOTES

1. The authors are deeply indebted to several individuals for reviewing and commenting upon this chapter, particularly Lillian H. Hoddeson and Nick Holonyak, both of the University of Illinois, as well as Ian Ross and William F. Brinkman, former president and current vice president for physical research, respectively, of the Bell Telephone Laboratories. An excellent, complementary account of the birth of the transistor appears in the book *Crystal Fire* by M. Riordan and L. H. Hoddeson (New York: Norton, 1997), a publication sponsored by the Sloan Foundation. We are also indebted to Probir K. B. Bondyopadhyay for extended discussions as well as access to valuable source material in his files. A detailed personal account of the developments that transpired during the period covered in this chapter, and that led to the invention of the bipolar transistor, is given in William Shockley's paper "The Path to the Conception of the Junction Transistor," *IEEE Transactions on Electron Devices*, vol. ED-23, no. 7 (1976); see also W. B. Shockley, *Electrons and Holes in Semiconductors* (New York: Van Nostrand, 1950); Sidney Millman, ed., *A History of Engineering and Science in the Bell System (1925–1980)* (Short Hills, N.J.: AT&T Bell Telephone Laboratories, 1983); Lillian H. Hoddeson, "The Discovery of the Point-Contact Transistor," *Historical Studies in the Physical Sciences* 12 (1981): 41; idem, "Research on Crystal Rectifiers during World War II and the Invention of the Transistor," *History and Technology* 11 (1994): 121. A biography of W. B. Shockley appears in the *Biographical Memoirs of the National Academy of Sciences,* vol. 68 (1996), p. 305.

2. Biographical material concerning Bardeen appears in *Physics Today* 45, no. 4 (Apr. 1992), a commemorative issue organized by David Pines. The authors are particularly indebted to Nick Holonyak for providing valuable information in the form of discussion and documentation concerning the critically important work of Bardeen and Brattain. This includes the privilege of viewing a lengthy televised copyrighted interview with Bardeen made in the United States in 1990 by Japanese television NHK. We are also indebted to John M. Anderson for a copy of an audio tape of a lecture Bardeen gave to a meeting of the Antique Wireless Association, Canandaigua, New York, Sept. 27, 1986, in which he reviewed the details of the early work. See also Hoddeson, "The Discovery of the Point-Contact Transistor."

3. In their book *Crystal Fire,* M. Riordan and L. H. Hoddeson mention failed attempts by J. Becker and W. Brattain to develop a working field-effect transistor in the mid-1930s with the use of cuprous oxide. Shockley repeated their attempt with Brattain's help later in the decade. Berthold Bosch in an unpublished essay, "Der Werdegang des Transistors 1929–1994, Bekanntes und Weniger Bekanntes" (an address given Nov. 17, 1994), describes several early attempts to develop a semiconductor triode. J. E. Lilienfeld, initially a member of the faculty of the University of Leipzig, clearly made a similar failed attempt a decade earlier. He migrated to the United States in 1927 and accepted an industrial position. See the biographical sketch of Lilienfeld by W. Sweet, "American Physical Society Establishes Major

Prize in Memory of Lilienfeld," *Physics Today* 41, no. 5 (May 1988): 87. Lilienfeld was subsequently granted three concept patents for devices that would closely resemble the present-day field-effect transistor, had it been possible to reduce them to practice: (1) J. E. Lilienfeld, "Method and apparatus for controlling electric current," U.S. Patent No. 1,745,175, filed Oct. 8, 1926, granted Jan. 28, 1930; (2) "Device for controlling electric currents," U.S. Patent No. 1,900,018, filed Mar. 28, 1928, granted Mar. 7, 1933; (3) "Amplifier for electric currents," U.S. Patent No. 1,877,140, filed Dec. 8, 1928, granted Sept. 13, 1932. See the paper by C. T. Sah, *Proceedings of the IEEE* 76 (1988): 1280. We are deeply indebted to Lillian H. Hoddeson and Probir K. B. Bondyopadhyay for detailed information on the patents of Lilienfeld. J. B. Johnson of the Bell Telephone Laboratories attempted to make working devices in accordance with Lilienfeld's patents and concluded that they were not viable. J. B. Johnson, *Physics Today* 37, no. 5 (May 1964): 60. See also Virgil E. Bottom, *Physics Today* 37, no. 2 (Feb. 1964): 24. There were, however, successful experiments carried out in France in the immediate postwar period (see page 174).

4. Bardeen's detailed analysis of the limitations imposed by surface states appears in a paper written during the early work on the transistor: J. Bardeen, "Surface States and Rectification at a Metal Semi-Conductor Boundary," *Physical Review* 71 (1947): 717. He concluded that they could have a significant influence on the properties of the contact potential if their density was 10^{12} per cm^2 or larger. See also the paper by W. E. Meyerhof, which follows this: "Contact Potential Difference in Silicon Crystal Rectifiers," ibid., p. 727. Apparently the first formal paper calling attention to the possibility of surface levels is that of I. Tamm, "Über eine mögliche Art der Elektronenbindungen an Kristalloberflächen," *Physikalische Zeitschrift der Sowjetunion* 1 (1932): 733

5. For information about G. L. Pearson, see J. Bardeen and W. H. Brattain, "Physical Principles Involved in Transistor Action," *Physical Review* 75 (1949): 1208.

6. The details of this and subsequent research leading to the invention of the bipolar point-contact transistor are presented in Bardeen and Brattain, "Physical Principles Involved in Transistor Action." The experiments are also described in the audiotaped lecture Bardeen gave in 1986 to the Antique Wireless Association, and in the 1990 videotaped interview with Japanese television NHK.

7. W. Schottky and E. Spenke, *Wissenschaft Veroff. Siemens Werken* 18 (1939): 225.

8. See Millman, *History of Engineering and Science;* also see J. H. Scaff, "The Role of Metallurgy in the Technology of Semiconductors," *Metallurgical Transactions* 1 (1970): 562.

9. Shockley, *Electrons and Holes in Semiconductors,* p. 59.

10. See Shockley, *Electrons and Holes in Semiconductors.*

11. J. N. Shive, *Physical Review* 75 (1949): 689.

12. See, for example, L. Hoddeson et al., eds., *Out of the Crystal Maze* (New York: Oxford University Press, 1992), p. 470.

13. For a very lucid account of several basic forms of junction transistor, see R. G. Hibberd, *Solid-State Electronics,* Texas Instruments Electronics Series (New York:

McGraw-Hill, 1968), lessons 5 and 7. By that publication date, the uses of the junction transistor had reached maturity. The field-effect transistor was beginning to play a major role.

14. Patent: J. Bardeen, "Three electrode circuit element using semiconducting materials," U.S. Patent No. 2,254,033, filed Feb. 26, 1948, granted Oct. 3, 1950.

15. Videotaped interview with Bardeen, Japanese television NHK, 1990.

16. W. Shockley, "A Unipolar Field Effect Transistor," *Proceedings of the IRE* 40 (Nov. 1952): 1365; G. C. Dacey and I. M. Ross, "The Field Effect Transistor," *Bell System Technical Journal* 34 (Nov. 1955): 1149; idem, "Unipolar Field Effect Transistor," *Proceedings of the IRE* 41 (Aug. 1953): 970. Field-effect transistors were the subject of much study and development in the 1960s and 1970s. Summaries of such work are given in the following two books in the Texas Instruments Electronics Series commissioned by Texas Instruments and published by McGraw-Hill: R. H. Crawford, *MOSFET in Circuit Design* (1967); W. N. Carr and J. P. Mize, *MOS/LSI Design and Application* (1972).

17. The development of the program for producing single-crystal wafers of silicon is reviewed, along with other material, in the following sources: Gordon Teal, W. R. Runyan, K. E. Bean, and H. R. Huff, in *Semiconductor Materials and Processing (Part A Materials)*, ed. J. F. Young and R. S. Shane (New York: Marcel Dekker, 1985); H. R. Huff and R. K. Goodall, "Silicon Materials and Metrology: Critical Concepts for Optimal IC Performance in the Gigabit Era," in *Semiconductor Characterization*, ed. W. M. Bullis, D. G. Seiler, and A. C. Seibold (New York: American Institute of Physics Press, 1996). See also the books on semiconductor technology in the Texas Instruments Electronics Series.

18. Doubtless, Mervin Kelly saw clearly the limitations of vacuum tubes very early since they were widely used in relatively isolated repeater stations in the telephone network. One of the quips commonly heard at the Bell Laboratories prior to the invention of the transistor was: "Nature abhors a vacuum tube!"

19. See the essay by Patrick E. Haggerty, *Management Philosophies and Practices of Texas Instruments* (Dallas: Texas Instruments, 1965); reprinted from *Proceedings of the IEEE* (Dec. 1964). This is reproduced in major part in the present book in appendix A. We are indebted to S. T. Harris and to J. Ross Macdonald, onetime director of research of Texas Instruments, for many discussions of the early history of the company in relation to its involvement with semiconductors.

20. For a broad, semipopular review of the development of the microprocessor, see, for example, Michael S. Malone, *The Microprocessor* (Santa Clara, Calif.: Telos, Springer, 1995). Among other matters, this book provides a special focus on the Intel Corporation.

BARDEEN AND SHOCKLEY:

New Careers

$\mathcal{T}$HE CLOSE PARTNERSHIP that Bardeen and Shockley had shared during the early period of work on the semiconducting triode, starting in 1945, did not endure into the 1950s. Apart from great differences in personality, Bardeen had, essentially since graduate school days, been interested in the existence of superconductivity in metals and had decided that he would devote a major portion of his research career to an attempt to discover its origin. While the Bell Laboratories would have been more than glad to have him stay on and work on the problem there, he decided that a position in a university in which he would have graduate students, postdoctoral fellows, and others as colleagues would be much more favorable for his program than the more specialized environment in which he was then working.

His decision was hastened by the discovery at Rutgers University, led by Bernard Serin, that the superconducting transition temperature was sensitive to the isotopic composition of the specimen upon which measurements were made, thus confirming Bardeen's prior conclusion that low-temperature superconductivity of the type first found by Kamerlingh Onnes in 1911 is probably linked to coupling between the electrons and the lattice vibrations. Actually, good academic positions were at somewhat of a premium in the early 1950s. The large number of returned veterans of World War II who had decided to take advantage of the GI Bill of Rights for support of a college education had for the most part completed their studies and gone on to their careers. Moreover, the birthrate in the 1930s had been relatively low because of the Depression, so that incoming classes of freshmen were small in the early 1950s. Many universities were retrenching. Fortunately, Dean William L. Everitt of the College of Engineering of the University of Illinois, in which both the electrical engineering and physics departments were situated, was able to present Bardeen with a reasonable offer in which his time and interests were to be shared by both departments, as Bardeen desired.[1] He remained at the university for the rest of his brilliant career, a highly admired and respected member of the faculty. It was there in 1956 that he learned that he, Brattain, and Shockley had been awarded the Nobel Prize for the invention of the transistor; it was there in 1972 that he learned that he, Leon N. Cooper, and J. R. Schrieffer had been awarded the Nobel Prize for their theory of what is now termed low-temperature superconductivity, observed in a number of metals and compounds near or below liquid helium temperatures.

Soon after Bardeen had departed from Bell Laboratories, Kenneth G. McKay, who had also joined the laboratories immediately after the end of the war, was selected to lead a transistor-development team that would work more or less in parallel to Shockley's group in order to maintain breadth of effort. Walter Brattain remained at the Bell Laboratories until his retirement and then moved to his native state of Washington.

THE SHOCKLEY TRANSISTOR COMPANY

Shockley stayed on at the laboratories until 1955. While in a sense he became something of a celebrity, both professionally and publicly, he came to feel that his remuneration at the laboratories did not reflect his true worth and accomplishments and that he would fare much better in another setting. He wished to be in charge of an independently created company devoted to research and development of transistors, about which he had many ideas. Having lived in California for most of his early life, he decided to return to that state in 1955 and seek entrepreneurial support.

While driving west, he spent a period in Urbana-Champaign to see old friends at the University of Illinois, and to negotiate by telephone from there with the various investment groups in California that were willing to support him in getting started. He finally agreed to accept financial backing from Arnold O. Beckman and established the Shockley Transistor Corporation in Palo Alto, a community he had known as a teenager. He became the president of the Shockley Transistor Corporation and director of the Shockley Semiconductor Laboratory. He had little difficulty in attracting a young, brilliantly creative staff since it was now amply clear that a new age in the field of electronics, still relatively undefined in detail, was emerging. Although he was not joined by any of his former colleagues at the Bell Telephone Laboratories, he remained in close touch with many of them in order to exchange information on detailed technical matters and to keep abreast of worldwide developments.

Paradoxically, Shockley's venture was both an outstanding success and an unfortunate failure. The best individuals he attracted received excellent indoctrination in the field from him but found him temperamentally difficult to work with. Apparently, he felt himself to be in competition with some of his own staff, particularly the physicists. Moreover, he decided that the company should focus its main attention on research devoted to the generation of valuable patents rather than to the profitable manufacture of cutting-edge devices. Since some of the staff understandably had more ambitious goals, the leading members of the group took their training with them to create other organizations nearby. Thus began the vast expansion of semiconductor enterprises in what is now known as California's Silicon Valley, an expansion represented in the diagram on pages 191–94. Bell Telephone Laboratories is represented by the lone box (under year 47) in the upper left-hand corner of the diagram. Shockley's company is just below it and to the right (under year 55). The other companies are spinoffs from his, independent startups attracted by the special environment, or the result of consolidations. Shockley, however, can truly be regarded as the Moses of that development.

Once the trend of events became clear to both Shockley and the sponsors of the new company, there was agreement that changes were in order. The company was eventually sold to the Clevite Corporation with Shockley serving as a consultant. On arriving in California, he had accepted a lectureship at Stanford University, a position that became a tenured professorship in the early 1960s.

Shockley barely survived a very serious, head-on automobile collision in 1961. It is difficult to say to what extent the accident changed his focus toward life, but after gaining essentially complete physical recovery he decided to devote the major part of his attention to eugenics issues, particularly those related to what he regarded as significant differences in the level of intelligence of various ethnic groups—all on his own terms. As a result, he achieved great notoriety and disrupted old friendships, which, on the surface at least, did not seem to bother him. Underneath it all, however, he may have suffered deep frustration.

A number of individuals who rose to prominence and fortune in the Silicon Valley owe something in the nature of a debt to Shockley. Among the most successful were Gordon E. Moore, a native Californian, and Robert N. Noyce, a midwesterner who had joined Shockley from a technical post in

Philadelphia in 1956. The two left with six others to form Fairchild Semiconductor, a division of Fairchild Camera and Instrument Company, and later,
in 1968, they went on to found the highly successful Intel Corporation.[2]
Noyce was to play an important role in the development of the integrated
circuit, as will be discussed in chapter 17.

John Bardeen, J. Robert Schrieffer, and Leon N. Cooper *(left to right)* at the Nobel award ceremony in Stockholm in December 1972. They had finally solved the riddle of the type of superconductivity of metals discovered by Kamerlingh Onness in 1911 at liquid helium temperatures. (Courtesy of David Pines and the University of Illinois.)

Kenneth G. McKay, who led a transistor development team that worked in parallel with Shockley's group after Bardeen's departure from the Bell Laboratories. (Property of AT&T Archives. Reprinted with permission of AT&T.)

William Shockley *(at the head of the table)* and members of the staff of Shockley Transistor celebrating the award of his Nobel Prize in November 1956. (Courtesy of SEMATECH.)

(Following three pages) The growth of the semiconductor-based industry in Silicon Valley from the founding of Shockley's company in 1955 until 1987. (Courtesy of Semiconductor Equipment and Materials International, Mountain View, California.)

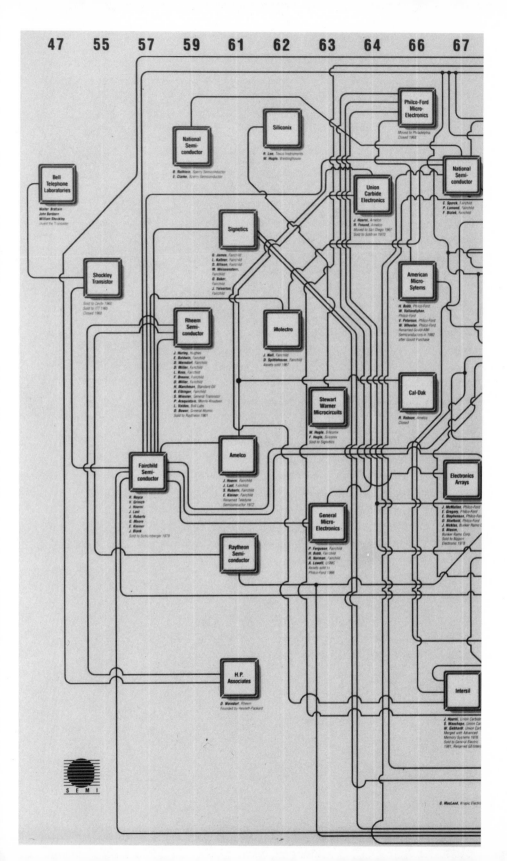

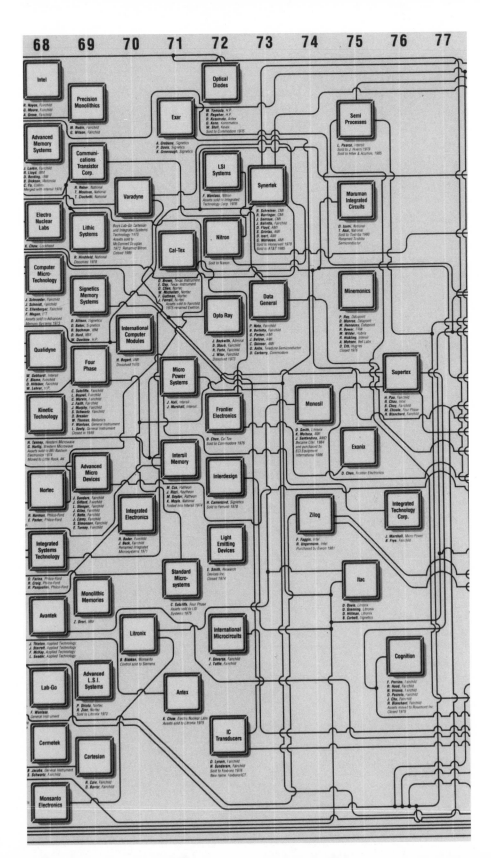

68   **69**   **70**   **71**   **72**   **73**   **74**   **75**   **76**   **77**

Intel
R. Noyce, Fairchild
G. Moore, Fairchild
A. Grove, Fairchild

Precision Monolithics
H. Rudin, Fairchild
G. Wilson, Fairchild

Optical Diodes

Exar
A. Grebene, Signetics
P. Davis, Signetics
K. Greenough, Signetics

Semi Processes
L. Pearce, Intersil
Sold to J. Hoerni 1979
Sold to Itten & Acurnos, 1985

Advanced Memory Systems
J. Larkin, Fairchild
M. Lloyd, IBM
D. Barding, IBM
B. Dickson, Motorola
C. Fa, Collins
Merged with Intersil 1976

Communications Transistor Corp.
N. Reber, National
T. Montoux, National
T. Ciochetti, National

Varadyne

LSI Systems
F. Wanlass, Nitron
Assets sold to Integrated Technology Corp. 1976

Synertek
R. Schreiner, CMI
R. Barringer, CMI
J. Santoux, CMI
D. Floyd, AMI
D. Grietas, AMI
B. Isert, AMI
G. Werleson, AMI
Sold to Honeywell 1979
Sold to AT&T 1985

Maruman Integrated Circuits
D. Izumi, National
T. Asai, National
Sold to Toshiba 1980
Renamed Toshiba Semiconductor

Electro Nuclear Labs
K. Chow, Lockheed

Lithic Systems
R. Hirshfeld, National
Dissolved 1976

Cal-Tex

Nitron
Sold to Natron

Computer Micro-Technology
J. Schroeder, Fairchild
J. Schmidt, Fairchild
G. Ellenburger, Fairchild
F. Megan, ITT
Assets sold to Advanced Memory Systems 1972

Signetics Memory Systems
D. Allison, Signetics
D. Baker, Signetics
R. Buchman, IBM
B. Hord, IBM
W. Davidow, H.P.

International Computer Modules
H. Bogert, AMI
Dissolved 1970

Opto Ray
J. Beckwith, Admiral
D. Staub, Fairchild
G. Parker, AMI
J. Belove, AMI
C. Skinner, AMI
G. Antle, Teledyne Semiconductor
D. Carberry, Commodore

Data General
R. Moto, Fairchild
R. Berletta, Fairchild

Minemonics
P. Ray, Datapoint
D. Monroe, Datapoint
M. Hennessy, Datapoint
R. Bower, TRW
M. Wilder, Hybrix
R. Hickling, Intersil
H. Mohsen, Bell Labs
D. Erb, Hughes
Closed 1976

Qualidyne
W. Gebhardt, Intersil
E. Blume, Fairchild
B. Hilleber, Fairchild
W. Lehrer, H.P.

Four Phase
C. Sutcliffe, Fairchild
L. Boysel, Fairchild
C. Marvin, Fairchild
D. Faith, Fairchild
J. Murphy, Fairchild
D. Schwartz, Fairchild
D. Breaker
W. Thomas, Melonics
M. Wanlass, General Instrument
J. Seely, General Instrument
Closed in 1985

Micro Power Systems
J. Hall, Intersil
J. Marshall, Intersil

Supertex
H. Pao, Fairchild
N. Chao, Intel
B. Choy, Fairchild
W. Chestle, Four Phase
R. Blanchard, Fairchild

Kinetic Technology
H. Tenney, Western Microwave
G. Hurlig, Western Microwave
Assets sold to GEI Baicom Electronics 1974
Moved to Little Rock, AK

Frontier Electronics
D. Chen, Cal Tex
Sold to Commodore 1976

Monosil
D. Smith, Litronix
K. Mutsou, AMI
H. Santandrea, AMI
Became Cibic 1984
and purchased by ECI Equipment International 1986

Exonix
D. Chen, Frontier Electronics

Nortec
J. Sanders, Fairchild
J. Gifford, Fairchild
J. Glenger, Fairchild
J. Giles, Fairchild
F. Botts, Fairchild
J. Carry, Fairchild
J. Simonsen, Fairchild
E. Turney, Fairchild

Advanced Micro Devices

Intersil Memory

Interdesign
M. Cox, Raytheon
J. Rizzi, Raytheon
M. Snyder, Raytheon
K. Moyle, National
Folded into Intersil 1974

Zilog
F. Faggin, Intel
R. Ungermann, Intel
Purchased by Exxon 1981

Integrated Technology Corp.
J. Marshall, Micro Power
B. Frye, Fairchild

Integrated Electronics
B. Bader, Fairchild
J. Beck, Fairchild
Renamed Integrated Microsystems 1971

Light Emitting Devices
L. Smith, Research Devices Inc.
Closed 1974

Itac
D. Davis, Litronix
D. Grenning, Litronix
B. Hillman, Litronix
B. Corbett, Signetics

Integrated Systems Technology
D. Farina, Philco-Ford
R. Craig, Philco-Ford
R. Pasqualini, Philco-Ford

Monolithic Memories
Z. Breri, IBM

Standard Microsystems
C. Sutcliffe, Four Phase
Assets sold to LSI Systems 1975

Avantek
J. Thielen, Applied Technology
J. Sterrett, Applied Technology
F. McKay, Applied Technology
L. Seader, Applied Technology

Litronix
B. Blakkan, Monsanto
Control sold to Siemens

International Microcircuits
F. Diverso, Fairchild
J. Tuttle, Fairchild

Cognition
F. Perrine, Fairchild
R. Hood, Fairchild
N. Vronis, Fairchild
D. Pezzolo, Fairchild
J. Cho, Fairchild
R. Blanchard, Fairchild
Assets moved to Rosemont Inc.
Closed 1979

Lab-Go
F. Wanlass, General Instrument

Advanced L.S.I. Systems
P. Shiota, Nortec
R. Zonn, No-Tec
Sold to Litronix 1973

Antex
K. Chow, Electro Nuclear Labs
Assets sold to Litronix 1975

Cermetek

Cartesian
R. Cole, Fairchild
D. Barter, Fairchild

IC Transducers
D. Lynam, Fairchild
N. Sunderam, Fairchild
Sold to Foxboro 1976
New name: Foxboro/ICT

B. Jacobs, General Instrument
B. Schwartz, Fairchild

Monsanto Electronics

LSI Systems (Cal-Tex)
D. Brown, Texas Instrument
L. Day, Texas Instrument
D. Chen, Nortec
W. Michaelsen, Nortec
F. Guttman, Nortec
K. Ferrell, No-Tec
Assets sold to Fairchild 1975 re-named Exertion

Buys Lab-Go Cartesian and Integrated Systems Technology 1970
Assets sold to McDonnell Douglas 1972 Renamed Nitron Folded 1985

H. Comentind, Signetics
Sold to Fairchild 1978

H. Yamada, H.P.
R. Reopher, H.P.
R. Kasumoto, Antex
G. Kano, Commatics
W. Shall, Kesels
Sold to Commodore 1976

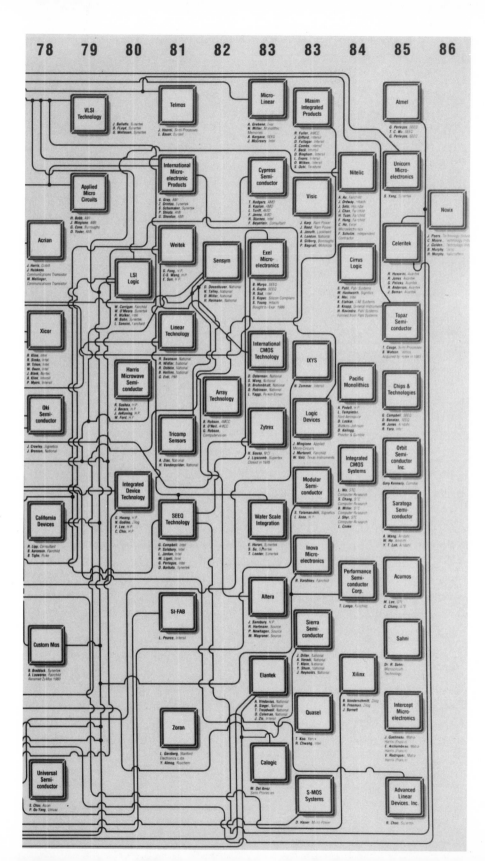

The group of eight members of Shockley's organization who left to form Fairchild Semiconductors. *Left to right:* Gordon Moore, Sheldon Roberts, Eugene Kleiner, Robert Noyce, Victor Grinich, Julius Blank, Jean Hoerni, and Jay Last. (Courtesy of SEMATECH.)

NOTES

1. See the discussion in F. Seitz, *On the Frontier: My Life in Science* (New York: American Institute of Physics, 1994).

2. The events that led to the creation of the Intel Corporation are discussed in Gordon E. Moore, *Daedelus* 125, no. 2 (1996): 55; see also the memorial book *Robert N. Noyce, 1927–1990* (SEMATECH, 1991), issued following Noyce's untimely death to commemorate his service to the semiconductor industry; Tom Wolfe, "The Tinkerings of Robert Noyce," *Esquire*, Sept. 1983. See also the interview with Gordon E. Moore, "Principia Moore," *Interface* 6, no. 1 (1997): 18; this gives an account of Moore's outlook on the prospects for the field as viewed in the early days of Fairchild Semiconductor, that is, circa 1959.

SIXTEEN

THE DEVELOPMENT OF TECHNOLOGY
AND LOGIC, 1948–60

$\mathcal{B}$EFORE REVIEWING the invention of the integrated circuit and its eventual evolution into the microprocessor, we must consider the hard-won technical advances that followed the invention of the discrete transistor.[1]

The large international group that took up the challenge started with little more than basic ideas. Silicon and germanium were now in a sense commonplace materials as a result of the earlier work on diodes. However, the steps and standards needed to produce reliable junction transistors in suitable quantity were essentially unknown, as were the forms of logic that would be needed to produce physical circuits that would use them most effectively.

The subsequent developments, which required intense, creative applications of chemistry and metallurgy, imaginative engineering planning and logical design, and further advances in solid-state physics, were guided by the groundwork that had been laid in two major areas of electrical engineering.

First, the transistor could be looked upon as a replacement for, or supplement to, the vacuum tube for applications that are said to be "analog" (or "linear") in nature. This includes, for example, the transmission and reception of conventional radio and television in which a band of frequencies linked to a carrier frequency carries the actual representation of sound waves or of appropriately scanned visual material.

Second, there were applications to what might be termed the programming of discrete logical processes such as switching or counting of electrical pulses. Both areas of engineering were already very important, but the second was gaining additional prominence because of the emergence of binary-digital programmed electronic computers, such as the ENIAC, developed at the University of Pennsylvania during World War II by J. Presper Eckert, Herman Goldstine, and John W. Mauchly, and computers of the Turing–von Neumann design, which were under construction at several places.[2] In fact, working versions such as the Ordvac and the Illiac were about to come on line at the University of Illinois. These early electronic digital computers, which were appropriately regarded to be marvels at their time, employed hundreds, and in some cases thousands, of vacuum tubes in their circuits. As a result, they not only consumed excessive power merely to heat their electron-emitting cathodes, but they suffered from the fact that they had relatively short lifetimes before burnout.[3] It was only natural to wonder if the transistor would prove to be a far more satisfactory substitute and to pursue its development.

LOGICAL ELEMENTS

A great deal of thought had already gone into the types of circuit elements that would be needed in electronic computers using the binary number system of logic. The three basic ones, or "gates," which are required, with appropriate interconnections that depend upon the overall plan of the system, are the OR, AND, and NOT circuits. The first has two or more inputs and one output. It emits a pulse if any one of the inputs receives a pulse. The AND gate also has several inputs and only one output. It, however, produces an output signal only when all the inputs receive one. The NOT gate has only one input and one output. It has the property that the output is the opposite of the input so that it is sometimes called an "inverter." Actually the NOT and AND gates may, for convenience of purpose, be combined in a single unit; the same is true for the NOT and OR gates.

It should be added that steps in the operation of a digital computer are activated simultaneously under signals derived from a digital clock, usually governed by a circuit that is tuned to a quartz piezoelectric-controlled oscillator.

ROLES OF GERMANIUM AND SILICON

Germanium was the obvious semiconductor of choice during the earliest period of research and development. It was easily available in the quantities needed at the time since good sources had been found during World War II; it has a substantially lower melting point than silicon (958° C rather than 1412° C); it could be refined to a reasonably high degree and produced in single crystal form with the method of zone processing developed by William Pfann in 1952 (see diagram, page 204). Although the surface oxide is water soluble, the hydrophilic qualities could be contravened to a substantial degree by encapsulating the units in a hydrophobic gel or wax.

Silicon was by no means neglected in view of the major role it had played in silicon point-contact diodes used as heterodyne mixers, displaying both ruggedness and relative insensitivity to temperature.[4] Moreover, the early work on transistors made it clear that a well-formed oxide layer on its surface possesses very special properties as an electrical and chemical barrier, unlike the oxide of germanium. It did not come into its own for use in transistors, however, until material of much higher purity than that used previously was available in single-crystal form.

Early in the 1950s a group at Texas Instruments decided that silicon should be the semiconductor of choice. Up to that point it had been used primarily in polycrystalline form, with the disadvantages that crystal boundaries and related imperfections (see page 181) offered as traps when minority carriers cross the base layer of the transistor. As a result, Gordon K. Teal, who had previously been at the Bell Laboratories and worked there on crystal growing with a colleague, J. B. Little, gained substantial support at Texas Instruments for a program designed to grow single crystals of silicon. The techniques for growing crystals had received a great deal of attention earlier in the century from both exploratory and applied scientists who were interested in the chemical, structural, and physical properties of crystals, particularly in cases where good natural specimens were not available.

After some testing, the Texas team settled on the so-called Czochralski method developed by J. Czochralski at the end of World War I and with which Teal and Little had experimented successfully at the Bell Laboratories. This method uses a rotating rod that possesses a thermal gradient and has a seed crystal of appropriate orientation attached to its end (see diagram, page 205). The rod is lowered into a molten pool of the material to be crystallized, in this case molten silicon, and is then slowly withdrawn while being rotated. The system proved to be highly effective, once numerous difficulties were overcome, including the selection of suitable materials for crucibles. It is widely employed at the present time to produce single-crystal ingots twenty centimeters in diameter, with thirty-centimeter ones in limited quantity. (See illustrations on page 249 for present-day examples.) This development caused somewhat of a sensation within the trade in its day since it meant that a new level of control of both the chemical and physical qualities of silicon had been achieved and that there would be greater standardization of product. Moreover, the focus of interest in germanium would abate since it lacked some of the most important virtues of silicon, including a very stable oxide. New, more refined, standards of quality in relation to what has come to be called "materials characterization," with all its connotations with respect to chemistry and physics, were dawning.[5]

Soon thereafter (1954), Texas Instruments marketed a light, silicon-transistor, battery-operated, frequency-modulation radio to indicate that wider commercialization of semiconductor devices was at hand. Later analysis showed that the sales, although substantial, were only marginally profitable as a result of setting too low a price. However a new era was definitely on the way.

FABRICATING TRANSISTORS

One of the earliest methods of preparing junction transistors used a fairly direct diffusion technique. Layers of a metal alloy containing a diffusable chemical element that could induce either n-type or p-type characteristics in the semiconductor were welded to opposite sides of a slab of semiconductor of opposite characteristic type. The unit was then heated to permit some of the added element in the alloy to diffuse into the semiconductor, thereby creating two p-n junctions. The metal layers on the surfaces of the unit could serve as bonding pads for leads to the emitter and collector. The portion of the original semiconducting slab formed the base. (See figures at the end of this chapter.)

Relatively elementary starts such as this were followed by a lengthy period covering most of the 1950s in which increasingly refined techniques derived from physical, inorganic, and organic chemistry, metallurgy, and crystallography were employed, often with applications of somewhat unrelated technology that had been used previously for other purposes. The single crystals of primary semiconductor were grown with ever-increasing attention to chemical content, including the levels of desired additives or "doping" agents as well as undesirable chemical elements that could act as traps or inhibitors for the migration of carriers. As mentioned in chapter 4, requisite quantitative standards eventually reached the range of parts per billion and better, with extensions of chemical spectroscopy providing the solution to chemical analysis.

Soon after the invention of the bipolar transistor in the 1940s, it was realized that imperfections in the arrangement of atoms in the crystal lattice could interfere with the operation of a transistor by trapping carriers and impeding current flow between the emitter and the collector. As a result, the full array of techniques for studying such imperfections that had been developed and used by crystallographers was introduced into manufacturing technology to detect them, often with major refinements made possible by the use of specially designed equipment.[6]

Layers of semiconducting material of one composition were deposited upon those of another by evaporation and sublimation, preferably so as to retain lattice continuity (epitaxial growth). Metallic layers intended as contacts for electrical leads were formed by evaporation or sputtering from an electric arc. Silicon dioxide layers were formed by dry oxidation methods at selected locations to serve as electric insulation or as shields against contaminants that could create surface traps for carriers. A layer of silicon dioxide could also prevent diffusion of dopant elements into the area it covered.

MASKING AND THE USE OF PHOTORESISTIVE MATERIALS

Masks that overlay the specimens and determined where the deposited material could or could not go were developed with increasing care and refinement as miniaturization advanced. These procedures were eventually supplemented with the use of layers of so-called photoresistive polymers developed by the Kodak Company, that is, organic materials that polymerize to a relatively insoluble form when exposed to appropriate wavelengths of light. The unexposed areas of the polymer, as determined by the design of the mask employed in conjunction with them, could be dissolved to expose the underlying silicon area where dopant could be selectively introduced, or a layer consisting of a metal, semiconductor, or oxide layer could be deposited.

Actually, the technology associated with the use of such photoresistive polymers had been developed first at the Diamond Ordnance Fuse Laboratory of the army as part of a broad miniaturization-integration program that was something in the nature of an outgrowth of the wartime work on the proximity-fused missile, as we shall see in the next chapter.

ION IMPLANTATION

In a more-advanced period, the 1960s, the group at the Bell Laboratories, which had installed high-voltage equipment in order to carry out experiments in nuclear physics and chemistry, found that it was practical to introduce controlled amounts of additional agents, such as elements from the third and fifth columns of the periodic chart, into silicon by ion bombardment. This technique of ion implantation has found a useful place in standard fabrication procedures.

MECHANIZATION

The mechanization of many of the manufacturing steps moved steadily apace with all these developments. While labor costs may have played a role to a degree in the initial trend, another motivation lay in the fact that instrumental control of mechanized or automated systems eventually provided greater yields of qualified product. It is probably safe to say that as the mechanized manufacturing procedures evolved, they eventually, in the 1970s and beyond, achieved heights unmatched in any other industry. A modern integrated chip fabricating unit, or "Fab," is a major achievement of chemical engineering.

DIVERSITY OF ACTIVITY: MESA AND PLANAR TRANSISTORS

As might be expected, a great deal of research was going on in many laboratories all through the decade of the 1950s in order to develop methods of producing individual transistors of appropriate design and composition. This included studies of ways of incorporating the various structures and additives that determined the nature of the carriers in the various portions of the devices.

Attention during this phase of activity focused on two principal types of junction transistor, known as "mesa" and "planar." The first, with double p-n junctions as described above, grew directly out of the geometrical structure developed by Shockley after the discoveries made by Bardeen and Brattain; the second, invented at Fairchild Semiconductor by Jean Hoerni, an associate of Robert Noyce and colleagues, was based on the realization that circuit elements could be spread out in the same plane and bridged by semiconductors or by metal leads and contacts, as circumstances require, and that elements could be insulated from one another with deposits of silicon oxide when desirable. The design is suggestive of an integrated circuit (see chapter 17), if one is prepared to take the additional step proposed by Kilby, namely, to make all the basic components of devices of silicon or germanium.

Overarching all of this, however, at the end of the 1950s, was the emergence of a revolutionary new development, the actual invention and development of the integrated circuit.

Morgan Sparks, of the Bell Telephone Laboratories, who made the first bipolar diffused transistor employing germanium. (Property of AT&T Archives. Reprinted with permission of AT&T.)

William G. Pfann, who developed the technique of zone refining. It served a very useful purpose in the initial purification of germanium. (Property of AT&T Archives. Reprinted with permission of AT&T.)

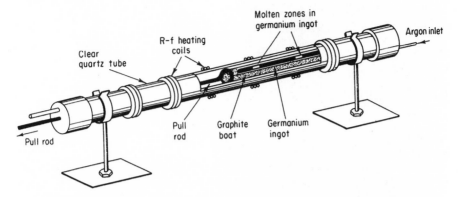

Schematic view of William Pfann's equipment for zone refining germanium. A boat containing the polycrystalline semiconductor is passed through successive heating zones in which the local temperature is sufficiently high to produce melting. After passing such a zone, the material freezes until it reaches the next heating zone. This method works well when the impurities present concentrate preferentially in the molten phase so that one gains successive stages of purification with each step of melting and freezing. (Courtesy of Robert G. Hibberd. Reprinted by permission of Texas Instruments Incorporated.)

William Pfann and Jack Scaff observing the operation of the zone refining equipment. (See also above.) Scaff, in the foreground, is holding a large single crystal of germanium made by the process. (Property of AT&T Archives. Reprinted with permission of AT&T.)

Gordon K. Teal, who succeeded in growing single crystals of purified silicon, a major advance in semiconductor and transistor technology. He found the Czochralski method (see below) to be the most satisfactory among several available ones. (Courtesy of the American Institute of Physics Emilio Segrè Visual Archives, Physics Today Collection.)

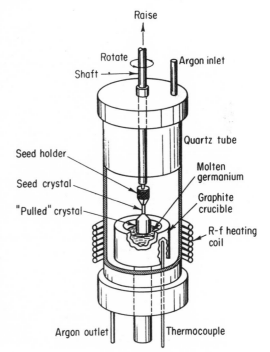

Schematic diagram illustrating the Czochralski method for growing single crystals. A rotating rod that has a selectively oriented seed crystal at its tip is lowered into a container of purified molten silicon. The rotating rod is slowly withdrawn and the thermal conditions adjusted to produce a long crystalline cylinder of controlled diameter from which slices, or wafers, can be cut. (Courtesy of Robert G. Hibberd. Reprinted by permission of Texas Instruments Incorporated.)

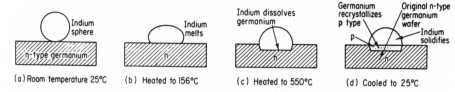

(a) Room temperature 25°C   (b) Heated to 156°C   (c) Heated to 550°C   (d) Cooled to 25°C

A schematic view of a method used in the early production of germanium transistors to incorporate indium into n-type germanium in order to obtain a p-n junction. (Courtesy of Robert G. Hibberd. Reprinted by permission of Texas Instruments Incorporated.)

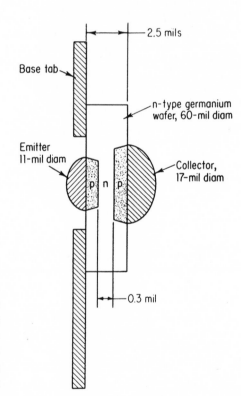

Schematic view of an entire germanium transistor created by the alloy-diffusion technique described on page 200. (Courtesy of Robert G. Hibberd. Reprinted by permission of Texas Instruments Incorporated.)

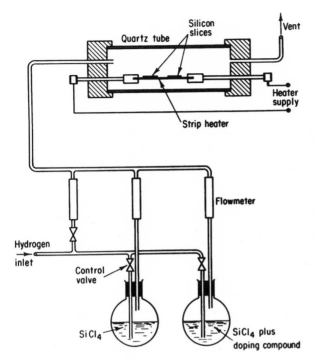

Schematic view of equipment for growing epitaxial layers of controlled composition on crystals of silicon. In the case shown, silicon tetrachloride vapor containing a predetermined amount of doping agent is reacted with hydrogen gas in a chamber (at the top of the diagram) in which slices of silicon have been placed. (Courtesy of Robert G. Hibberd. Reprinted by permission of Texas Instruments Incorporated.)

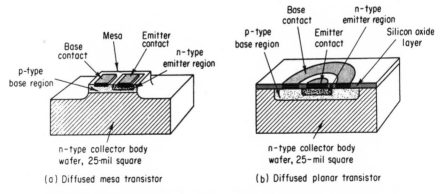

Two successively advanced designs of discrete diffused silicon transistors. Both types are mounted on the collector body. In the left-hand device, the base region forms a mound, or *mesa,* on top of the collector and the emitter is formed by diffusion into the base. In the right-hand planar device, the base is formed by diffusion into the plane of the collector and the emitter by diffusion into the plane of the base. (Courtesy of Robert G. Hibberd. Reprinted by permission of Texas Instruments Incorporated.)

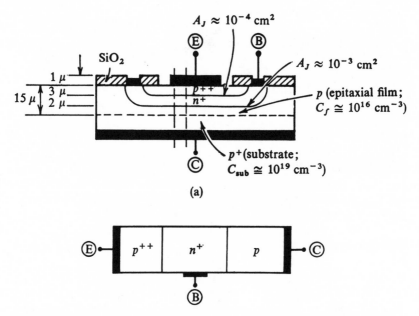

(a)

Cross section of an advanced discrete planar transistor, giving dimensions and levels of doping in the various segments. The emitter, base, and collector are identified by capital letters in a circle, the doping levels in two portions of the collector by capital C with a subscript. The symbol A_j is accompanied by the value of the width of the junction. The diagram at the bottom shows schematically the basic structure of a bipolar junction transistor. (Courtesy of Andrew S. Grove of the Intel Corporation. Derived from his book *The Physics and Technology of Semiconducting Devices* [New York: Wiley, 1967]. Reprinted by permission of John Wiley and Sons, Inc.)

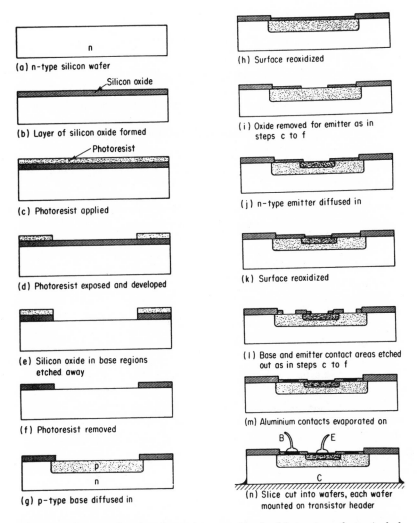

(a) n-type silicon wafer

(b) Layer of silicon oxide formed

(c) Photoresist applied

(d) Photoresist exposed and developed

(e) Silicon oxide in base regions etched away

(f) Photoresist removed

(g) p-type base diffused in

(h) Surface reoxidized

(i) Oxide removed for emitter as in steps c to f

(j) n-type emitter diffused in

(k) Surface reoxidized

(l) Base and emitter contact areas etched out as in steps c to f

(m) Aluminium contacts evaporated on

(n) Slice cut into wafers, each wafer mounted on transistor header

Schematic diagram showing the various steps in the fabrication of a typical planar transistor. (Courtesy of Robert G. Hibberd. Reprinted by permission of Texas Instruments Incorporated.)

NOTES

1. For background, see the Texas Instruments Electronics Series, commissioned by Texas Instruments and published by McGraw-Hill; Sidney Millman, ed., *A History of Engineering and Science in the Bell System (1925–1980)* (Short Hills, N.J.: AT&T Bell Telephone Laboratories, 1983); J. Bardeen and W. H. Brattain, "Physical Principles Involved in Transistor Action," *Physical Review* 75 (1949): 1208; A. S. Grove, *Physics and Technology of Semiconducting Devices* (New York: Wiley, 1967). See also *Materials Science and Technology*, ed. R. W. Cahn, P. Haasen, and E. J. Kramer, vol. 16, *Processing of Semiconductors*, ed. K. A. Jackson (New York: VCH Publications, 1996). An excellent account of special aspects of the technical problems encountered in developing semiconductor electronics is given in the essays in a forthcoming book by A. Goldstein and W. Aspray, *Facets: New Perspectives on the History of Semiconductors* (IEEE Center for the History of Electrical Engineering). We are grateful to the authors for the privilege of seeing an advance copy of the manuscript.

2. In keeping with the principle of parallel international invention, a German engineer, Konrad Zuse, started developing a sequence of binary computers in the mid-1930s, working in comparative isolation. He achieved the "first fully automated, program-controlled and freely programmable computer using binary floating-point calculation in 1941." He employed electromechanical relays. A close associate urged him to shift to vacuum tubes, but the cost and the power consumption inhibited him from turning his relatively small enterprise in that direction. He was astonished at the end of the war to learn of the size of the ENIAC, which employed vacuum tubes. See Zuse's autobiography (in English), *The Computer—My Life* (Heidelberg: Springer, 1991). We are indebted to Helmut Völcker of the University of Essen for a copy of the book.

3. There is a story, perhaps apocryphal, to the effect that Enrico Fermi doubted that the electronic computer had much future, when first proposed, because of the limitations of the vacuum tube. Incidentally, Fermi exhibited considerable skepticism concerning the value of research in solid-state chemistry and physics during World War II. While at the University of Chicago laboratories devoted to the production of plutonium and studying the possible damage to reactor components that might result from the high level of radiation to be expected, Frederick Seitz requested the addition to the group of an outstanding experimental chemical physicist, Robert Maurer, who was experienced in the details of solid-state chemistry and physics. Fermi, who was in nominal charge of the overall research program, demurred, saying, in effect, that the field was overemphasized. Fortunately, Eugene Wigner intervened; the addition to the staff rendered heroic service to the program. Unfortunately, Fermi did not live to see the revolution wrought by semiconductor technology.

4. See R. G. Hibberd, *Solid-State Electronics*, Texas Instruments Electronics Series (New York: McGraw-Hill, 1968).

5. An excellent summary of the work at the Bell Telephone Laboratories devoted to materials characterization is given in Millman, *History of Engineering and Science.*

6. Accounts of the struggles over the control of crystal imperfections, with particular emphasis on dislocations, are presented in the following essays: H. R. Huff and R. K. Randal, "Challenges and Opportunities for Dislocation-Free Silicon Wafer Fabrication and Thermal Processing: An Historical Review," forthcoming in the *Conference Proceedings of the Third International Rapid Thermal Processing Conference,* ed. R. B. Fair and B. Lojek (the conference was held in 1995); H. R. Huff and R. K. Goodall, "Material and Metrology Challenges for the Transition to 300 mm Wafers," *Conference Proceedings on 300 mm Wafer Technologies and Materials* (article based on a paper presented at the January 1996 SEMICON conference, held in Korea); Don Rose, "Future of 200 mm Wafers," *Proceedings of SEMICON/West 1993.*

SEVENTEEN

THE INTEGRATED CIRCUIT

𝒯HE DEVELOPMENT of proximity fused weapons during World War II required that their electronic circuits be as compact as was feasible. They were designed to explode in grenade-like fashion when the reflection of a signal emitted by an internal, vacuum-tube, radio transmitter indicated that they were approaching the target. The success of the electronic part of the development ultimately led to the compaction or integration of other electronic devices in which the components were tightly clustered on a ceramic base. The initial work was carried out in the Central Laboratory Division of the Global Union Corporation, under the sponsorship of the National Bureau of Standards (now the National Institute for Science and Technology) and with the key guidance of Merle Tuve of the Carnegie Institution in Washington, D.C., mentioned in chapter 6 in connection with studies of the ionosphere with radio waves.

After the war, the work on integration was greatly extended for applications such as hearing aids and other devices. In the process of fabrication, mechanized methods of assembly that reduced costs substantially were also developed. The invention and widespread use of printed circuit boards that provide a convenient means for mounting and connecting electronic components emerged from this important phase of development.

Once the discrete transistor began to be a standard member of the electronics family, it was inevitable that it would become involved in the advances in integration of the systems in which it was employed. Fairly early in this cycle (1952), G. W. A. Dummar of the Royal Radar Establishment in Britain made the following prediction: "It seems now possible to envisage electronic equipment in a solid block with no connecting wires. The block may consist of layers of insulating, conducting, rectifying and amplifying materials, the electrical junctions connected directly by cutting out areas of the various layers."[1] It was left to others to fulfill this goal, the pioneers being

Jack S. Kilby, of Texas Instruments, and Robert N. Noyce, then at Fairchild Semiconductor.

KILBY'S INVENTION

Jack Kilby has written a vivid, fair-minded account of the two approaches in which he explains how he recognized that a novel concept, appropriate for the new medium, was needed.[2] As usual, it required specially prepared and insightful individuals to appreciate the opportunity and be sufficiently inspired to achieve the goal of integration. Following the war, Kilby had been employed in semiconductor research and development at the Central Laboratories of Global Union, but he decided in 1958 to transfer to a larger company dedicated to the full-scale manufacture, as well as the use, of semiconductors. He selected Texas Instruments.

Newly hired, Kilby did not have the privilege of a leave at the firm's first vacation break, which occurred soon after his arrival. Undisturbed by the normal level of activity, he decided to use the time at work to develop his own thoughts on what might ultimately prove to be the best way to achieve integration. He conceived the idea of having the semiconducting substrate, with engraved transistors, resistors, capacitors, and interconnects, as the complete integrated circuit—essentially all components being fabricated from the wafer of germanium or silicon. The management of the company, not least Patrick Haggerty, was highly enthusiastic about the concept, and the work went ahead rapidly after researchers had produced working circuits that demonstrated the basic soundness of the concept. It was announced publicly at the March 1959 meeting of the Institution of Radio Engineers. In the meantime, patent applications had been developed and filed.

Fortunately for the program, far-seeing individuals at the Wright Air Development Command of the U.S. Air Force saw the potential virtues of the technology in weapons systems and provided much-needed financial support during the 1960s. At first, there was substantial objection to the entire concept from some conservative members of the engineering profession outside the company, who feared that it would be difficult to fabricate adequate resistors, capacitors, and substitutes for conventional inductors from silicon. Time and experience eventually demonstrated, however, that the advantages that resulted, when suitable compromises were made, greatly outweighed the objections. In particular, phase-shift circuits served well in the place of conventional inductors. The road to the future of the technology was open, provided appropriate resources could be mustered.

FAIRCHILD DEVELOPMENT AND ROBERT NOYCE

At about the same time, and partly as a result of the announcements that emerged from Texas Instruments, other organizations subsequently filed for patents complementary with Kilby's concepts. Most significantly, Robert Noyce and his colleagues at Fairchild Semiconductor, who had been developing planar discrete transistors (see chapter 16), realized along the way that it might be possible, in principle at least, to build a number into a single slice of semiconductor—a natural basis for a patent. The planar approach relied much on the use of the durable, insulating, silicon dioxide that can be thermally grown or deposited on the surface of silicon.

Turner Hasty, a physicist who was associated with both Kilby and Noyce in connection with two different assignments in his career, is familiar with much of the background history of both inventions. Hasty recalls that Noyce had his vision at about the same time as Kilby but was first inclined, more or less in the spirit that had caused Mervin Kelly to "free up" the basic patents for the transistor, to leave the concept open and available for widespread development. Hasty has added that he thinks that of the two highly talented inventors, Kilby initially had a far broader general vision of what might lie ahead.

PATENT NEGOTIATIONS

The matter of patent rights fell into the hands of the lawyers and law courts. Texas Instruments persisted in claiming priority and eventually gained a strong patent position in the 1960s. Among many other rewards, the company used its patent strength to acquire the right to construct ownership operations, including manufacturing facilities, in Japan. Indeed, the Japanese media used the words "Texas Instruments" as a term of opprobrium for a period of time—at least until Japanese companies gained strong market share in the field of semiconductor electronics. The profession and the public have given comparable and richly deserved recognition to both Kilby and Noyce for their work over the course of time.

MEMORY STORAGE

One major consequence of the invention of the integrated circuit was a significant revolution in data storage, or memory. In the post–World War II period of electromechanical computers, punched cards were commonly used for the storage of information. This was, for example, the case in equipment sold by the International Business Machines Company (IBM) in its earlier days.

A shift to magnetic storage began in the 1950s, once magnetic systems became reliable and inexpensive as a result of the development of polymeric materials that could form an adequately strong and relatively inexpensive supporting matrix for finely divided magnetic materials. Several forms of magnetic storage were used for different purposes. For a decade or longer, large reels of magnetic tape, which were limited in speed and access, served a very useful purpose for both random access memory (RAM) and read-only (ROM) memory when retrieval time was not a factor or there was no reasonable alternative. Such tape had actually been developed in Germany in the 1930s, but it had been reserved for Hitler's private use in connection with his radio speeches. The technology became open and it advanced rapidly after the war, particularly in the hands of the Ampex Corporation in Redwood City, California, which took advantage of the German development.

The pigments department of the DuPont Company, which had played a critical role in the production of relatively pure silicon, contributed a magnetic form of chromium oxide to the development of magnetic tape of superior quality.

One variant of magnetic recording, which was developed at the Bell Telephone Laboratories and had a relatively short life, consisted of a high-speed, rotating magnetic drum on which information could be stored, read, or removed by means of an associated induction coil that traversed the length of the drum on a sliding mechanism, making it possible to gain access to any portion of the drum fairly rapidly.

The Massachusetts Institute of Technology invented and developed another system that was very popular throughout the 1960s. This system employed an interconnected network of small toroidal (doughnut-shaped) magnets that could be magnetized in either direction along the axis of rotation of the toroid. Wires that were capable of transmitting current pulses that could reverse the direction of magnetization of the toroids, or that could "read" the state of their magnetization, were threaded through the openings in the toroids. In the late 1960s, the Ampex Company constructed a three-dimensional lattice of this type of storage unit involving the use of one million such magnets.

STORAGE ON INTEGRATED CHIPS

While magnetic memories, as represented, for example, by both hard and "floppy" discs, continue to play a very important role in the storage of information, particularly for relatively long-term or archival purposes, the development of the integrated circuit, particularly those based on the field-effect transistor, made it possible to create chips containing huge capacity for storing data in both read/write and read-only memories.[3] Both types are usually configured in an architecture that permits random rather than series access. The random access read/write memory (RAM) is available in two forms, static random access memory (SRAM) and dynamic random access memory (DRAM). Static memory cells embody a configuration of four or six transistor flop-flop circuits for information storage, whereas the dynamic cell stores information as charge on a capacitor. The DRAM cell consists of one capacitor and one transistor.

The information stored in the SRAM remains in the system until it is erased or the power is turned off. On the other hand, the stored charge on the DRAM cell will leak off with the passage of time, requiring a clock-determined refresh cycle, in addition to the normal read and write cycles. Both types of memory are volatile in the sense that neither can retain stored information when the power is removed. The DRAM, having the fewest components per cell, has the obvious advantage of packing density, and hence cost per bit. The SRAM has the advantage of speed since no refresh cycle is needed. In general the largest available DRAM will be about four times the size of the SRAM.

Read-only memory (ROM) also comes in two forms, the factory-programmed ROM and the field-programmed ROM (PROM). The ROM is frequently a component of a larger integrated circuit such as a microcontroller, microprocessor, or digital signal processor. It is usually stored with operating instructions for the chip.

At present, the sixteen-megabit DRAM is generally available in the market and the sixty-four–megabit form is becoming more readily available. Currently all such large-scale memory chips are fabricated using complementary MOS (CMOS) technology with devices having submicron features. Bipolar static RAM is limited to rather small sizes because of power consumption. It has only a few military applications.

The first semiconductor memory chips appeared in the early 1970s and contained of the order of one (1972) and four (1974) kilobits of binary storage. With the rapid domestic and international evolution of technology, however, this had grown to sixty-four kilobits by the end of the decade.

CHARGE-COUPLED DEVICES

The development of the integrated circuit has also made it possible to create a special form of memory called the charge-coupled device (CCD). To date, at least, it has found only specialized uses, mainly for the sensing of visual images, particularly in hand-held video cameras.

Viewed at the fundamental level, it consists of a linear sequence, or string, of depletion potential wells, the depth and shape of which can be varied during the course of operation. A minority charge can be introduced into the well at one end of the linear sequence and passed along from one well in the line to the next, to be withdrawn and recorded at the opposite end. The stepwise transit of the charge from one well to the next in the series is mandated by changes in the depth and shape of the wells in a synchronous way, so that the entire line of charges is shifted at once. When storing digital data, the charge introduced at the starting end is either zero or a fixed, unit amount. Parallel, two-dimensional arrays of such storage registers can be used to store a two-dimensional matrix of numbers.

If relatively long-term storage is desired, the charge that reaches the terminal end of a line can be reintroduced at the starting end so that the line serves as something like an endless storage belt for preserving the sequence until it is called upon to serve the demands of a processor.

A two-dimensional CCD array can also be used to store and read out the photoelectrons (or holes) that are produced in the substrate by an optical or X-ray image and captured in the wells. Among other applications, such devices have proven to serve as very sensitive recorders in astronomical research. They are employed at low temperatures in order to preserve the stored charge for the period of an extended measurement. The use of low temperatures can also serve to reduce the background "noise" associated with thermal fluctuations within the storage system.

Jack S. Kilby, who conceived of the integrated circuit in 1958 shortly after joining Texas Instruments Incorporated from the Central Laboratories of Global Union, where he had been involved in more conventional integration of circuits. (Courtesy of Jack S. Kilby.)

Robert N. Noyce, who conceived of the integrated circuit at about the same time as Kilby and after he and his colleagues created Fairchild Semiconductor Company. He was strongly influenced by the work on the fabrication of planar, discrete transistors. (Courtesy of the Intel Corporation.)

NOTES

1. G. W. A. Dummar of the Royal Radar Establishment, in an address to the Electronics Component Congress in 1952; quoted in J. S. Kilby, "Invention of the Integrated Circuit," *IEEE Transactions on Electronic Devices* 23 (1976): 648.

2. See Kilby, "Invention of the Integrated Circuit."

3. We are indebted to Turner Hasty for a detailed account of the current status of memory storage on integrated chips.

EIGHTEEN

ADVANCES IN THE 1960S AND
VISIONARY FORECASTS

$\mathcal{T}$HE DECADE of the 1960s proved to be a peculiarly complex one for the evolution of silicon electronics, providing the final bridge between the old and the new eras. The proposal of entirely new design and fabrication procedures for electronic systems by Kilby and Noyce opened up fascinating vistas of a new world. The challenges that stood in the way of achieving that world, however, were enormous. Far more than capital would be needed, although it was clearly an absolutely essential ingredient. Foremost, moving ahead would require the guidance of unusually farsighted, inspired, and capable leadership, backed by the restless, diligent creative work of many scientists and engineers, most of whom would be inspired by their own vision of the future. They would not only have to solve a large number of difficult technical problems, but would, along the way, be required to identify and articulate the problems that would have to be solved. Fortunately, the rewards to be anticipated from success were sufficiently great to provide all of the incentive necessary for those who chose to become involved, including the major sponsors, which, initially at least, were the federal agencies.

Both Fairchild and Texas Instruments possessed at that time exactly the type of exceptional leadership and dedicated staff needed to face the challenges. The companies were competitors, but in a climate in which competition provided incentives for innovation. Rivalry eventually worked to the benefit of both. Moreover, many other organizations, such as AT&T, IBM, RCA, and Motorola, were contributing to the advance of the technology with their own interests in mind.

THE INSIGHTS OF HAGGERTY AND MOORE
(1964 AND 1965)

In a paper first published in the December 1964 *Proceedings of the IEEE,* Patrick Haggerty, the remarkably imaginative and articulate head of Texas Instruments, described with much eloquence and clarity the situation faced by the electronics industry in the mid-1960s. Gordon Moore, then at Fairchild Semiconductor, proffered his own highly cogent forecasts a year later, in the pages of *Electronics* in October 1965. One cannot do better than to consider some of their observations, starting with those of Haggerty. The essentially complete texts by Haggerty and Moore are given in appendixes A and B, respectively, in this volume. Indeed, we are fortunate to have these exceptional, forward-looking documents available. Although Haggerty writes from his professional origin as an electrical engineer, and Moore, in a sense, from his as a research chemist, their statements flow together in a remarkable way.

HAGGERTY'S VIEWS

Among the highlights to be noted as Haggerty explores his visions at such an early date is the prediction that, if the vexing technical problems related to reliability and containment of fabrication costs are overcome, the new electronics will completely permeate the activities of society, causing a major revolution in the way things are done. Still further, the advances in the technology, including increases in complexity of circuit design and computing power, will in turn facilitate the further advance of reliability and complexity through something in the nature of a feedback mechanism. The new technology will provide major aid for its own development. Finally, if the immediately pressing technical problems related to manufacture are solved, the cost per active electronic group will drop substantially and the volume of production will increase. On the latter point, Haggerty turned out to be much too conservative. His projections of costs ten years later (1974) were much too high and his estimate of production volume too low. It must be recalled, however, that in 1964 the scientists and engineers involved in production faced great difficulties in achieving commercially practical yields of units. Taken as a whole, Haggerty's insights turned out to be remarkably prescient.

GORDON E. MOORE AND MOORE'S FIRST LAW (1965)

In 1965, a year after Haggerty had written his farsighted essay, Gordon E. Moore, one of the founders of Fairchild Semiconductor Company, director of its research and development laboratories, and later a founder of the Intel Corporation, followed with another highly remarkable series of forecasts.[1] Indeed, he was willing to be more courageous with respect to forecasts of costs and volumes of product. Much progress had been made in advancing the technology during the intervening year and it was beginning to be amply clear that a ready market, still mainly—but not entirely—for governmental defense agencies, was at hand. In the essay, which appears in appendix B of the present book, Moore enunciated the principles that came to be known as "Moore's law" (later, Moore's first law), a challenging speculation at the time:

> The complexity for minimum component costs has increased at a rate of roughly a factor of two per year. Certainly over the short term this rate can be expected to continue, if not increase. Over the longer term, the rate of increase is a bit more uncertain, although there is no reason to believe it will not remain nearly constant for at least 10 years. That means by 1975, the number of components per integrated circuit for minimum cost will be 65,000.
> I believe that such a large circuit can be built on a single wafer.

COMMENTARY

Although Moore was prepared to accept the state of the art as it stood in 1965 as highly advanced and essentially well established, the overall tenor of his essay indicates clearly that he believed the technology was still in its early stages. Indeed the development of the microprocessor, another revolutionary step, lay just ahead.

To those interested in such comparisons, Moore has pointed out in an essay written for *Daedelus* in 1996 that the cost of producing an acre of silicon chips has remained at about one billion dollars since the time of the first commercial silicon transistor.[2]

TRANSITION TO DIGITAL PROCESSING OF SEISMIC DATA

A fine example of the manner in which the new digital-based semiconductor technology actually led to very substantial improvements in the handling of analog seismic data is given in a historical account of the experience at Texas Instruments prepared by Harvey G. Cragon of the University of Texas at Austin.[3] The work described by Cragon represents one of the first large-scale applications of the new technology and is dramatized by the successive improvements achieved with the use of successive generations of digital computers.

THE CREATION OF INTEL

In 1968, Robert Noyce and Gordon Moore concluded that the time had come to form a company of their own. They were highly experienced leaders in the field of integrated circuits, which had proven to be fully capable of providing the avenue for the future development of electronics. They had gained a great deal of experience in manufacturing and sales and were now aware of the vast number of programs that could be profitably developed if most of the resources produced by a company were plowed back to support further development. They sensed that Fairchild Semiconductor was only one of the significant interests of those who led the parent Fairchild enterprise and that they could do much better on their own. Moreover, the parent company not only seemed to be facing an internal crisis in selecting a new leader for its semiconductor company, but seemed almost permanently committed to power-consuming bipolar technology whereas the field-effect devices, which were relatively economical of power, seemed more appropriately adaptable for very large memory chips. The result was the creation of the Intel Corporation, which, like Texas Instruments under Haggerty, quickly became one of the leading companies in the field.

With the departure of Noyce and Moore, many other members of Fairchild Semiconductor sought other positions or started companies of their own. Fairchild filled the numerous vacancies by acquiring new staff, many from the Motorola Corporation, which had established its semiconductor center in Phoenix, Arizona. Included in the group from Motorola was C. Lester Hogan, the head of the Phoenix operation, who had once been at the Bell Telephone Laboratories. Fairchild continued to do well, and was eventually sold, first to the Schlumberger Corporation and then, in 1988, to National Semiconductor Corporation, led by Charles Sporck at the time.

Intel started off at a fast pace with the development of large-capacity semiconductor memory chips, filling a major vacuum in the memory field, which had previously been served by interlinked magnetic core devices. The company began with bipolar systems but shifted to several types based on the use of metal-oxide-semiconductors, gaining a long lead on the competition until it became necessary to adopt a new strategy in the 1980s. The new Intel development was soon profitable; the company developed rapidly.

Noyce served as head of the company until 1975 and then became chairman of the board. Moore succeeded him as chief executive.

PROLIFERATION

Although much attention has been devoted to Fairchild and Texas Instruments in this chapter because of their leadership roles, it must be emphasized that individuals in many companies, both large and small, were engaged in research and development concerning transistors and their applications. In fact the interest in the field was growing internationally. One gains the impression of a gathering army of participants who improved older concepts, provided new inventions and insights, and, in general, were doing their best to enhance the art.

Moreover, many companies both inside and outside the semiconductor field were beginning to establish assembly and fabrication plants, as well as distribution channels, in suitable countries abroad in order to take advantage of factors such as low costs of skilled or unskilled labor and growing cadres of well-educated professionals, and to establish a presence in growing markets. This marked one of the initial phases of what has come to be called the evolution of the world marketplace.

It is to be noted in passing that almost none of the large companies that had played major roles in the development of vacuum tube technology in the decades prior to World War II did so in a comparable way during the evolution of transistor technology. In this connection, a solid-state physicist at the General Electric Laboratory, LeRoy Apker, who had an excellent background in inorganic and physical chemistry, was, in the 1950s, assigned to help the groups in the company that were developing the early versions of transistors to standardize their manufacturing procedures. Many of those involved were individuals who had previously been engaged in research and manufacturing of vacuum tubes. Others were employed in divisions devoted to the production of filament or fluorescent lamps, which had to face declining, or at least changing, prospects. Apker stated that his most difficult task was to convince the workers that they needed to achieve new, much higher standards of chemical control in producing transistors than they had been accustomed to in their prior work. Establishing such standards would be much easier to attain in a new, enthusiastic, start-up organization containing a fresh, hand-picked manufacturing staff.

FIELD-EFFECT TRANSISTOR

As was mentioned earlier, considerable attention was being devoted to improving the field-effect transistor during the decade. This endeavor finally led to what became one of the major developments of transistor technology. By 1970 the so-called metal-oxide-semiconductor field-effect transistors (MOSFETs) were fully as important in many forms of service as the junction transistors that had offered the greatest promise initially.

Patrick E. Haggerty *(center)* in the early 1960s. He is flanked by J. Erik
Jonsson *(left)*, one of the founders of what is now Texas Instruments, and
Mark Shepherd *(right)*, who worked closely with Haggerty on the tran-
sistor program and succeeded him as chief executive officer in 1969.
(Courtesy of Texas Instruments Incorporated.)

Gordon E. Moore, who, as director
of research, joined the group of
eight to found Fairchild Semiconductor. He became chief executive officer of Intel when Robert
Noyce gave up that position in
1975. (Courtesy of Intel
Corporation.)

Robert Noyce and Gordon Moore, as founders of the Intel Corporation in the late 1960s. (Courtesy of SEMATECH.)

The triumvirate that led the development of Intel. *Left to right:* Andrew Grove, Robert Noyce, and Gordon Moore. Grove, who also has a background in advanced chemistry, is the current chairman of the board and chief executive officer. He followed Moore as chief executive officer in 1987, when the latter retired from the position, and recently became chairman of the board. Moore is now chairman emeritus. (Courtesy of Intel Corporation.)

NOTES

1. In addition to Moore's *Electronics* essay, see his article in *Daedelus* 125, no. 2 (1996): 55. Also relevant is the article by Phillip E. Ross, "Moore's Second Law," *Forbes*, Mar. 1995, p. 116.

2. See Moore's article in *Daedelus*.

3. H. G. Cragon, "The Early Days of the TMS 320 Family," *Texas Instruments Technical Journal* 13, no. 2 (1996).

NINETEEN

THE 1970S AND THE MICROCONTROLLER

*T*HE 1970S PROVIDED additional great advances in silicon technology, in remarkable fulfillment of the forecasts of Haggerty and Moore. Associated with the new decade was the prompt arrival of a major innovation that accelerated the revolution in electronics and enhanced its pervasive penetration into many aspects of everyday life, professional and otherwise. That innovation was the microcontroller, or microprocessor, anticipated in a sense but emerging almost completely by surprise. Both Intel and Texas Instruments claim credit for the invention, but, without insisting that its development was "inevitable," it seems safe to say that the microcontroller lay clearly on the pathway being opened by highly creative individuals in the most dynamic and innovative companies and was bound to emerge in their hands in the 1970s, sooner or later.

TEXAS INSTRUMENTS

Following the introduction of the integrated circuit, Texas Instruments moved ahead along a very broad front.[1] It was becoming a well-seasoned company in its new, main area of interest and was looking well beyond the federal agencies for new markets. In fact, it had become a major supplier of integrated circuits to many commercial customers, IBM being one. The immediate technical barriers that Haggerty had foreseen in 1964 had, for the most part, been overcome. New ones had been encountered, but there was no reason to believe that they could not be surmounted by the highly motivated staff of the company. With so much success behind Texas Instruments and with such great promise for the future, some of the younger leaders in the company developed the slogan "We can do anything!" With this spirit in play, two aggressive developmental activities were undertaken. They are recalled by Gary Boone, a leader in the activity:

> The first of these was an original design, carried out in 1970–71, of a single-chip Central Processing Unit (CPU), or microprocessor (called

TMX1795), pursuant to a cathode ray tube terminal customer's original architecture and single-chip-ness specification requirements. It should be noted that both Intel's 8008 and Texas Instrument's TMX1795 were designed to meet the same customer specifications (a company once known as CTC, but now known as Datapoint). Thus, both these single-chip CPU designs executed substantially the same instruction set. A TMX1795 prototype worked about a year before an Intel 8008 did. Intel went on to production with the 8008, 8080, 8088, 8086, etc. processors. The TMX1795 never got into production.

The second project was an original architecture and original design, also carried out in 1970–71, of single-chip calculator or microcontroller chips (initially called TMS1802 and later designated TMS01XX, where XX digits denote mask-programmed variations). The TMS01XX project was directed to developing a versatile architecture to "program" with multiple calculation-oriented feature sets and thereby serve multiple customers' requirements, while achieving engineering and manufacturing economies associated with mask-programming variations of a common design. Several TMS01XX inventions were patented, showing the immediate calculational functionality, but also envisioning diverse applications such as odometers, taxi-fare meters, digital volt meters, counters and the like. Thus, TMS01XX inventions led the way for the self-contained processing power of microcontrollers (CPU, plus program, plus data, plus input, plus output, all on a single chip) becoming imbedded in literally billions of products. Today, more than 2,000,000,000 microcontrollers are consumed annually.

Subsequently, Texas Instruments developed the Datamath©, a battery-powered, palm size calculator that used the popular TMS0102 in the form it was designed to be used.[2]

This initial development was followed, along with much competition from outside, by watches, bilingual dictionaries, electronic games, and, eventually, desktop computers and evermore powerful hand-held calculators for scientific and business use. By the end of the 1970s, the most advanced computers based on the use of microcontroller technology were beginning to enter areas of application that had previously been the exclusive province of the manufacturers of mainframe computers, such as IBM, and minicomputers, such as those produced by Digital Equipment Corporation (DEC).

In June 1996, the United States Patent Office gave official recognition to the priority of Boone's invention of the microcontroller. This followed a long period of litigation as a result of conflicting claims. It will be noted that in his statement, given above, Boone distinguishes between what he terms "the microprocessor," which processes data within the computer, and "the microcontroller," which not only serves that function but also has a broader one from input to output. This usage has become somewhat blurred in more recent times, although it has been retained in the U.S. Patent Office.

EVENTS AT INTEL

In the case of Intel, it appears that there was an element of chance and opportunism.[3] A Japanese research engineer, Mashatoshi Shima, involved in research at a small Japanese desktop calculator company, Busicom, decided in 1969 that it ought to be possible to design and produce a relatively simple, transistorized version of the arithmetic calculator, employing no more than a few chips. His company put out a proposal for bids based initially on a design he developed. A new member of Intel's staff, Ted Hoff, who had previously gained experience on circuit design at Fairchild, examined Shima's design and proposed a simpler alternative. Initially, Shima decided to continue to work with his own design. The idea had taken root in Hoff's mind, however, and he gained management approval to develop his own design further, in spite of the fact that Intel was mainly looking for business linked to the manufacture and sale of high-volume, technically advanced units in immediate demand at that early stage of the company's development.

When Hoff showed a more advanced, but simpler, version of the computer to Shima, the two agreed to attempt to develop the concept together and eventually emerged with a four-chip design. The issue might have stalled there since it was by no means clear how best to organize the interconnections. Fortunately, the problem was, at this critical point, turned over to Federico Faggin, also newly recruited from Fairchild. He was an expert in such design logic, and, after experimenting, he achieved a workable solution that was transferred to Busicom in March 1971. The Japanese company was interested in the new system only for its uses in a calculator and willingly transferred other rights to Intel after negotiations. This was a very fortunate outcome for Intel since it was just beginning to dawn on the in-house staff there that the new system, with variations in design, had potential uses in a great many applications. It contained all of the basic elements of a full-fledged controller, including input and output ports, an arithmetic-logic unit, read-only and random access memories and timers. What was needed was more development, as well as challenges of application, once the flexibility and power of the new system began to be appreciated more widely outside the company. Advances in integrated circuit technology would guarantee ever-increasing capabilities and convenience of use.

BROADENING OF USE

Companies such as Motorola that, among other things, had strong, specialized, interests of a traditional nature, including mobile communications systems, soon found new uses for microprocessors in old areas, in this case in electronic controls of increasing sophistication for the operation of automobiles. Such applications opened up a worldwide market for controllers, particularly as pressures for economy of operation and minimum polluting emissions grew.

Perhaps more important than any other development has been the extent to which the growing capacity of microcontrollers, along with decreasing cost per unit circuit element, has facilitated both national and international trade. Barring the emergence of a great war or a great business depression, commercial interests alone should guarantee supporting pressure for the continued growth of that technology.

LIBRARY CATALOGS

In another more modest but useful direction, one of the distant dreams of the 1960s—to place the entire catalog of a substantial library into a computer in order to increase the speed of search and the geographical range of access—began to become feasible. This happened as microcontrollers evolved and the cost of semiconductor memories dropped dramatically, along with increases in the number of elements on an integrated circuit chip.

At the time this is written in the 1990s, the standard catalog information concerning the English-language collections in the Library of Congress acquired since 1968 (and in some cases earlier) has been made available to users of a telephone-linked personal computer system.

FABRICATION FACILITIES

As the demands for more complex chips grew, manufacturing technology began to be more highly mechanized and automated. Whereas an advanced fabrication facility was to a large extent a manual, even batch, operation in the late 1960s, as Haggerty forecast in his 1964 essay (see appendix A), by 1980 the most advanced production units were not only beginning to be highly automated, representing the highest form of the art of chemical engineering, but involved the presence of a minimum number of operating personnel in the areas where integrated circuits were being fabricated. The required standards of cleanliness exceed by far those encountered in any other industry, except possibly special portions of the pharmaceutical and medical device industry, for quite different reasons. Miniaturization proceeded apace and the geometrical resolution of structures placed on a chip approached and penetrated well below the so-called one-micron barrier, with a corresponding increase in the number of electronically active components per chip. Moore's law was holding steady.

As will be recalled, Haggerty thought in 1964 that the cost of an advanced fabrication facility would soon rise to a level of several million dollars and that only a few companies would be able to afford them if the market did not grow with sufficient rapidity. Actually the cost of such units climbed to levels of the order of thirty million dollars as the 1970s progressed, but the rapid growth of demand for microcontrollers and memories, along with increased efficiency of production, encouraged the continuous proliferation of fabrication units. The possibility of concentrating production in a few companies was not even a matter of further speculation until the 1990s.

JAPANESE COMPETITION

By the mid-1970s, the Japanese producers had taken full advantage of the patent rights to the integrated circuit they had obtained in their bargaining with Texas Instruments and were proving to be highly competitive in many important markets. Moreover they worked with the motto "Do it right the first time!" As a result, they succeeded in establishing a reputation for unusually high quality and reliability of product. They possessed another great advantage. The business of production and sales proved to be cyclical once the volumes increased. Periods of strong sales were followed by ones in which the market was saturated and there was excess capacity. It was the custom of companies in the United States to delay the construction of new, more advanced, fabrication units during periods of slowdown of demand in order to control cash flow as much as possible and to avoid diluting earnings, which would affect share price. The Japanese, however, had a type of federal support through the banking system that permitted them to expand facilities during such periods of slowdown so that they could take full advantage of the next period of high demand for more advanced devices. This practice earned them increasing market share in many product areas as the decade progressed.

One simple but good example of Japanese foresight and prowess is the approach to the problem of the well-known liquid crystal display. This device involves the selective absorption of a polarized component of unpolarized light by appropriately oriented long-chain molecules that can be aligned by an electric field. The RCA laboratory in Princeton, New Jersey, had carried out experiments that demonstrated the basic feasibility of generating a readable display from such materials, but it did not pursue the development to the point of achieving a practically usable product. Most electronic companies in the United States that explored the device as a result of the initial work at RCA did so in a relatively superficial way but decided not to push the matter further. The Japanese took the time and effort to explore this ultimately very useful display thoroughly and gained the advantage of a very high market share.

THE SEMICONDUCTOR INDUSTRY ASSOCIATION

As the decade advanced, Robert Noyce took one of the first major steps to bring some degree of order into the semiconductor industry in the United States with the hope of encouraging more standardization and coordination. In 1977, he obtained agreement from an initially small group of companies, of which Intel was one, to form the Semiconductor Industry Association. Undoubtedly, he hoped that the organization would encourage cross-industry discussions of mutual problems. He had become one of the leading statesmen of the industry.

Another organization, formed by the joint action of several semiconductor manufacturing companies, is SRC, based in the Research Triangle Park in North Carolina. It has provided a very valuable mechanism for channeling industrial funds to university laboratories to promote research and development, as well as the education of students, in the field of semiconductor technology. The results obtained are made generally available.

Gary Boone, who, while at Texas Instruments, led a team of about six engineers that developed the first microcontroller (also known today as the microprocessor) on a single chip of silicon. The device first operated successfully on July 4, 1971. The single-chip microcontroller, with suitable variations in design, has been used extensively, providing the platform for the evolution of the digital computer. (Courtesy of Gary Boone.)

Ted Hoff, who joined Intel from Fairchild Semiconductor in 1970. He responded to a request-for-bid from a Japanese electromechanical computer company, Busicom, for help in the design and manufacture of semiconductor circuitry for a transistorized computer. Along the way, Hoff realized that the underlying device had far greater potential use when redesigned as a general-purpose microprocessor. (Courtesy of Intel Corporation.)

Federico Faggin, who, as a new acquisition at Intel, worked with Hoff in
programming what became the
company's first microprocessor. He is
cofounder and currently president of
Synaptics Corporation in Silicon
Valley. (Courtesy of Synaptics
Corporation.)

NOTES

1. Much information concerning technical developments at Texas Instruments and elsewhere can be found in the books in the Texas Instruments Electronics Series, published by McGraw-Hill.

2. Letter, Gary Boone to Norman G. Einspruch, July 22, 1996.

3. See the account in Michael S. Malone, *The Microprocessor* (Santa Clara, Calif.: Telos, Springer, 1995); see also G. E. Moore, "Cramming More Components onto Integrated Circuits," *Electronics*, Apr. 19, 1965, p. 114, and idem, "Intel—Memories and the Microprocessor," *Daedelus* 125, no. 2 (1996): 55. Also see Phillip E. Ross, "Moore's Second Law," *Forbes*, Mar. 1995, p. 116.

TWENTY

1980–2000 AND THE FUTURE

𝒯HE PERIOD SINCE 1980 has been marked by as complete a fulfillment of the visions offered by Haggerty and Moore (see appendixes A and B) as one could expect. The pervasiveness of electronics has increased and the number of semiconductor devices manufactured has continued its exponential growth. There may be a period of readjustment and industrial reorganization just ahead, as a result of rising costs of fabrication facilities, but no immediate limitations arising from technical sources are apparent.

The deaths of Patrick Haggerty and Robert Noyce were major losses among farsighted industrial leaders during this period. The first died of pancreatic cancer in 1980 and the second of coronary fibrillation in 1990, both in their mid-sixties.

Haggerty's death was followed by a relatively smooth transition in leadership at Texas Instruments since Mark Shepherd and J. Fred Bucy, who were already highly experienced executives in the company, had already taken over the top posts. Nevertheless, there was an inevitable change in psychological atmosphere and the company encountered something of a mid-life crisis through a loss of market share in major products, a common phenomenon in an industry where fluctuations are the norm rather than the exception. Among other things, the staff had been slow to adopt field-effect technology for use in many devices in which it would have been most appropriate and paid a price in leadership at a critical time. The company resumed its stride under the guidance of Jerry R. Junkins, who was previously in charge of the very successful government electronics business. He took over as chief executive officer in 1985, at the time Bucy retired from that post, and during a recession. He introduced new business concepts that matched the requirements of the period. He served in office for eleven years, until his untimely death in 1996. He was succeeded by Thomas Engibous, an electrical engineer who had been in charge of the semiconductor electronics business.

237

In the meantime, Andrew Grove, who had been at Fairchild with Noyce and Moore, assumed the leadership role at Intel and is providing continuing dynamic and visionary guidance.[1] In 1997 he became chairman of the board, but retained his guiding role. Gordon Moore is now chairman emeritus.

SEMATECH

One of Noyce's last major contributions to the industry in his role as statesman was to accept the post of chief executive officer of SEMATECH. This is a consortium of fourteen member companies that joined together in 1988 under the guidance of the Defense Advanced Research Projects Agency of the Department of Defense to develop common policies with respect to the specification and evaluation of manufacturing equipment. The name of the organization is an acronym for Semiconductor Manufacturing Technology. Turner Hasty of Texas Instruments became Noyce's deputy. The formation of the organization was inspired, in part, by the fact that Japanese competition was threatening the existence of some of the most important domestic producers. Each domestic company had been selecting its own suppliers of manufacturing equipment with perhaps more emphasis on cost than quality. Someone in the new organization developed the slogan "The great secret about semiconductors is that there are no manufacturing secrets!" By standardizing on a set of common metrics that are used to evaluate and improve semiconductor equipment, and placing prime emphasis upon quality, the companies soon overcame their handicaps and regained worldwide competitive positions.

MOORE'S SECOND LAW (1996)

The decades since the seventies were also marked by intense intercompany competition, with great fluctuations from time to time in what customers regarded as leadership products. As fortunes alternated, the overall situation had some of the aspects of a complicated greyhound race in which advanced technology replaced the mechanical rabbit as the lure. Along with this, the cost of an advanced fabrication unit, a magnificent marvel of chemical engineering, has reached and exceeded the billion-dollar level, giving rise to what is sometimes called Moore's second law (see appendix B and graph on page 265), which implies that a major economic crisis is at hand throughout the industry. Actually, the situation is somewhat reminiscent of the early days of automobile manufacture, prior to 1930, when there were in the United States alone twenty or so manufacturers competing for the market. One would normally expect some form of consolidation to occur along with much excitement and some degree of trauma. In fact, Texas Instruments, under Junkins's leadership, formed investment partnerships with several Asian companies in the construction of fabrication facilities for common use. These are managed by Texas Instruments with commitments on the allocation of product output. Intel, however, remains in a sufficiently strong position to be able to construct multibillion-dollar fabrication units with its own resources. In another direction, Taiwan Semiconductor Manufacturing Company, one of the major producers of custom-made semiconductor components, including silicon wafers, is laying plans to expand its production several fold in a stepwise manner, using its own resources if the growth market justifies the investment.[2]

SOFTWARE

It was commonplace in the 1970s to express serious concern about the way in which circuit design—that is, the creation of complex software—would be managed in the long run as integrated circuits became ever more complex. Actually, matters have worked out surprisingly well. In effect, the evolving microprocessor came to its own rescue by playing a central role in the development of software through the process known as computer-aided design (CAD), much as Haggerty predicted in 1964. The growth in intricacy of the processor has made it possible to consolidate past gains in software into well-formulated subprograms so that the challenge to the designer of logic at any given time lies in making the next incremental gain. Even here, however, the designer has the advantage of access to powerful simulation and analytical equipment with which to test complex circuits fairly thoroughly. The final computer-based designs can then be used to create photomasks and guide the manufacturing process.

YIELDS

One major consequence of the use of computer-aided design is that the initial yields of acceptable product are much higher than they were previously. In the 1970s, one felt gratified to have an initial yield of 5 or 10 percent, with the expectation that it could be raised to a value closer to 30 percent by a struggle involving much trial and error. Usually one shifted design and manufacturing procedures to smaller, more efficient, circuit dimensions when that level of yield was reached. Today initial yields are frequently in the range of 80 percent (see page 247). The overall gross yields have also been raised by increasing the diameter of the wafers of silicon used in manufacturing integrated circuits. Twelve-inch wafers are being developed and should soon be in use (see page 249).

In the meantime, the geometrical dimensions of the structures that can be reproduced on chips have diminished to the range of a quarter micron, with the expectation that another factor of two in reduction can possibly be achieved by pushing the present form of optical lithography to the limit. Doubtless, there will be pressure to go further, and new techniques based on the use of X rays or electron beams could come into play. On the purely theoretical side, one might well expect to reduce linear dimensions by another factor of ten or so, encountering quantum effects along the way.[3] New challenges, involving complex chemistry, physics, and engineering, will arise but they will inevitably be linked to new opportunities.

THE SPECIAL ROLE OF SILICON

The properties of elemental silicon are so well suited for the role it has come to play that it is difficult to believe that the material will not remain central to developments in the foreseeable future.[4] Moreover, the cumulative investment in technological research and equipment for dealing with silicon and exploiting its use is now so large that it would be very costly to attempt to replace it as long as no unexpected obstacles arise. This does not mean that no attention will be given to other materials such as gallium arsenide and related compounds for companion use, usually in special applications, such as light-emitting diodes and electron optics. The evolution of materials to use in photodiodes and the development of quartz fiber optics provide good examples of parallel interests.

One significant limitation on the use of silicon emerges at the microwave level of frequencies. The intrinsic electrical conductivity of pure silicon is sufficiently high at normal operating temperatures that useful amplifiers are limited to the range below eight to ten gigahertz, that is, to wavelengths above three to four centimeters. The resistivity can be increased by alloying silicon with germanium. Devices made with the alloy can operate at wavelengths below one centimeter.

MICROPROCESSORS

The ceaseless growth of the power of the microprocessors since their inception in 1971 has continued to change the role of the mainframe processor. The mainframe still has an important place in the management, processing, and distribution of massive amounts of data, but it is frequently linked to a network of smaller processors that draw information from it for special purposes. Should the density of the active elements on a chip increase by another factor of fifty or so, as seems to be beyond question, the line separating mainframe and desktop processors may become even less sharply drawn.

LANGUAGE PROCESSING

One of the areas of basic and applied science that has benefited enormously from the advances in capacity and speed of computers is the field of language processing.[5] A significant part of the gain is associated with increases in the basic understanding of language, that is, the science of linguistics, and in the compression of messages made possible by conversion from analog to digital representation. Much of what has been accomplished would have been deemed almost beyond hope in the 1960s. It is still far from the time when one can expect to speak arbitrarily in one language and expect to have it appear with only minor errors in another, in either voice or print format, but some progress is being made in that direction in at least a limited way at the laboratory level. What is clearly in sight, if progress continues, is the equivalent of a voice-controlled word processor and its converse in a clearly usable form, although initial success may require some constraint in vocabulary and articulation as well as in sentence structure. This ability will carry with it the capacity to use relatively complex spoken orders to control complex devices, including microprocessors, and to obtain useful information over networks of stored information by verbal request.

In an interesting, forward-looking paper, Sadaoki Furni of the Nippon Telephone and Telegraph Company predicts optimistically that many significant breakthroughs in these areas at advanced, sophisticated levels will be achieved by the turn of this century.[6] Others, such as Stephen Levenson, are more cautious.[7]

CELLULAR AND MOLECULAR BIOLOGY

One of the noteworthy developments of recent decades has been the extent to which the biochemical community has adopted the electronic computer as a major general-purpose working tool. By coincidence, the growing complexity of electronic computers over the past forty years or so has paralleled closely the growth of understanding in the field of molecular biology. Moreover, the molecular biologists who are in the forefront of this development soon learned that the new technology could be of great use to them when modeling complex molecules in three dimensions, when recording complex data, such as lengthy sequences of molecules in proteins and genetic material, and in controlling complex experiments. The basic laboratory needs of the biochemist today are no less complex than those of the bench physicist. The marriage to the computer is a permanent one.

USER-FRIENDLY SYSTEMS

For the average person, one of the greatest gains in the development of processors has been the creation of what are called the user-friendly systems of tools for entry and operation. This area was opened up by the Apple Computer company in 1984 with the introduction of the Macintosh system. Whereas entry language was previously tied exclusively to operations on the keyboard with critically sensitive rules that had to be memorized, the new system makes extensive use of the video monitor, along with icons and other graphic aids as well as worded instructions. The system can be operated by means of an electronic pointer (cursor) linked to a movable "mouse."

The far-seeing individual who risked the fate of his company on this commercial venture was Steven Jobs, whose name became a household word as a result. He apparently was inspired to push the new development as a result of seeing an experimental model of a unit that contained the basic features, albeit in more primitive form, at the Xerox Research Laboratory in Palo Alto. The introduction of the new system was greatly aided by the expanding capacity of the microprocessor, which made it possible to provide the user with far more explicit guidance in avoiding errors and in correcting them when they did occur.

The Microsoft Company, created by William Gates, has now introduced versatile software that has the benefit of being compatible with the products of most other hardware providers. In any event, the end result of the development of such user-friendly systems has been the opening up of much increased use of the microprocessor for a very broad range of individuals, extending from elementary school children to business people.

THE INTERNET

The rapid proliferation of personal computers in the past two decades has made it feasible to develop the Internet system, which can link users both with one another and with various centers of information. The employment of the network is made particularly easy at present by the fact that normal telephone lines are adequate for most of the needs of the average person, although extended delays are becoming more and more frequent because of popular use of special programs as well as congestion of telephone circuits. The system, however, is still in its infancy, resembling in its own way the sequence of developments that occurred in the early days of radio and television.

The universal introduction of fiber-optic networks could alleviate many of the problems associated with limitations on the number of available channels by increasing that number many fold. The opportunities for diversity are almost unlimited. It remains to be seen what high levels of use, coupled with commercialization, will foster and whether the more open, public channels of the system will become cluttered and overwhelmed with trivia.

Patrick E. Haggerty *(left)*, in the mid-1970s. He is shown here on the campus of Rockefeller University in New York City with David Rockefeller. Haggerty was at that time serving as chairman of the board of both Texas Instruments (until 1976) and the university. (Courtesy of Rockefeller University.)

Robert N. Noyce, during the late 1980s when he was serving as chief executive officer of SEMATECH, a consortium of fourteen U.S. semiconductor companies established in Austin, Texas. Noyce had by this time become the principal spokesman for the industry. (Courtesy of Intel Corporation.)

Mark Shepherd, who took over the post of chief executive officer of Texas Instruments when Haggerty shifted his position to chairman of the board in 1969. Shepherd became chairman in 1976 and retired in 1988. (Courtesy of Texas Instruments Incorporated.)

Jerry R. Junkins, who followed J. Fred Bucy as chief executive officer of Texas Instruments in 1985. The company had faced difficulties in retaining its competitive position in the 1980s, but Junkins, who had spent many years with the company, reestablished its standing among the leaders. He served for eleven years before his untimely death in 1996. (Reprinted by permission of Texas Instruments Incorporated.)

Andrew S. Grove, who
became chief executive
officer of Intel in 1987
and helped continue the
upward course of the
organization. (Courtesy
of Intel Corporation.)

Turner Hasty, who became the repre-
sentative of Texas Instruments at
SEMATECH and held the position of
deputy to Robert Noyce while there. He
is typical of the knowledgeable veterans
of the semiconductor industry, having
witnessed and lived intimately with
most aspects of its growth.
(Courtesy of Turner Hasty.)

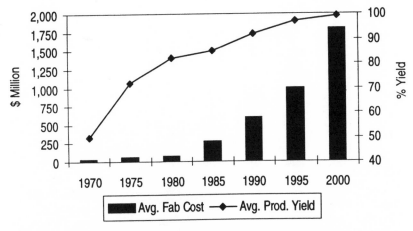

The improvement in percentage yield of acceptable devices per wafer of silicon as a function of years. Also shown is the rising cost of fabrication units, now in the range of two billion dollars. (Courtesy of Dan Klesken of Robertson Stephens and Company, San Francisco.)

Steven P. Jobs, who, as leader of the Apple Computer company, conceived of a user-friendly computer system, Macintosh, which became highly successful. (Courtesy of Eric Seitz, the Next Software Corporation, and Apple Computer, Inc.)

The Microsoft team that worked in Albuquerque with its leader, William Gates, as it was constituted in 1978. Gates is at the lower left in the photograph. The others are *(top row):* Steve Wood, Bob Wallace, Jim Lane; *(second row):* Bob O'Rear, Bob Greenberg, Marc McDonald, Gordon Letwin; *(front row):* Andrea Lewis, Marla Wood, Paul Allen. (Courtesy of the Microsoft Archives.)

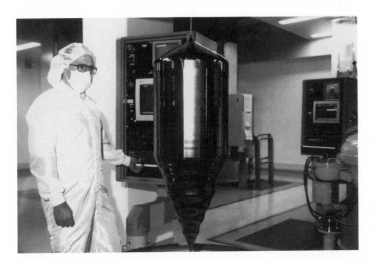

A single crystal ingot of silicon having a diameter of 30 centimeters (12 inches). This presumably will become a standard product in the near future as the drive to obtain evermore efficient production of integrated circuits continues. The weight of the entire crystal is 120 kg. (Courtesy of H. Fusstetter and Wacker Siltronic AE.)

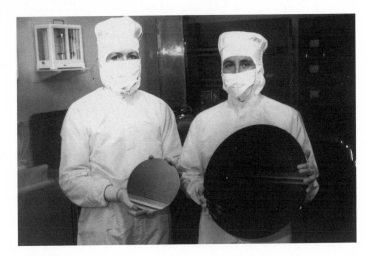

The shape of things to come. The two technicians are holding wafers of single crystal silicon. The wafer on the left, which has served as the standard during the 1990s, has a diameter of 20 centimeters (8 inches) and a thickness of 725 microns. The one on the right has a diameter of 40 centimeters (16 inches) and a thickness of 950 microns. (Courtesy of H. Fusstetter and Wacker Siltronic AE.)

NOTES

1. Grove has written a lively and highly informative book that deals with his own period of stewardship of the company: *Only the Paranoid Survive* (New York: Currency-Doubleday, 1996). Among many other things, the book contains an account of the way in which he and his colleagues dealt with a crisis that arose in the 1980s when Japanese companies captured the memory market, upon which Intel had depended for it profits. They shifted successfully to the production of microprocessors. At the time the book was written the company was searching to find its position with respect to the rapidly evolving Internet.

2. The remarkable growth of the semiconductor industry in Taiwan is described in the *Journal of Industry Studies* 3, no. 2 (1996) (a publication from the University of New South Wales Press, Sydney). We are indebted to Dr. K. T. Li for a copy of the relevant material.

3. A review of research on quantum effects related to electrical conductivity appears in H. van Houton and C. Beenakker, "Quantum Point Contacts," *Physics Today* 49, no. 7 (1996): 22.

4. A general review of the status of semiconducting materials appears in the article by H. R. Huff, "Semiconductors, Elemental—Material Properties," forthcoming in the *Encyclopedia of Applied Physics,* vol. 17 (New York: VCH Publishers). See also, by the same author, "Silicon Materials Science and Technology: A Personal Perspective," forthcoming in the *Proceedings of the 189th meeting of the Electrochemical Society, Incorporated* (the meeting was held in May 1996). An account of technological developments as they stood in 1990 is presented in W. R. Runyan and K. E. Bean, *Semiconductor Integrated Circuit Processing Technology* (Reading, Mass.: Addison-Wesley, 1990).

5. The status of this field as it stood in 1995 is described in a series of papers in the *Proceedings of the National Academy of Sciences,* vol. 92 (1995), pp. 9911ff.

6. Ibid.

7. Ibid., p. 9953.

PATRICK HAGGERTY'S
FORECAST (1964)

The following paragraphs are reproduced from the book by Patrick E. Haggerty, *Management Philosophies and Practices of Texas Instruments* (Dallas: Texas Instruments, 1965). They are restricted to the portion headed "Integrated Electronics—A Perspective" (pp. 123–35). This part first appeared in the *Proceedings of the IEEE* (December 1964). Subheadings have been added here, and several discursive paragraphs have been omitted.

IMPORTANCE OF INTEGRATION

The technologies which offer the largest promise for the achievement of integration in electronics today as well as those which seem to have the largest future potential are those which use a common substrate and continuous processing for forming a large proportion, if not all, of the circuit elements. Specifically, these technologies depend upon the use of thick (silk screen) and thin films and semiconductor techniques. In the growth of integrated circuits during the past five years and, indeed, in promise for the next decade, semiconductor technology is dominant.

Thus, the beginnings of integrated electronics, as we know it now, can be traced to World War II, the proximity fuse program, and techniques developed in 1945 by the National Bureau of Standards and Centralab, for forming resistance and capacitance on a ceramic substrate by silk screening of conductive inks. However, the key event signaling the advent of true integration in electronics was unquestionably the development during 1958 by Jack S. Kilby of Texas Instruments of a concept of processing the equivalent elements for a complete circuit, such as resistors, capacitors, transistors, and diodes, in a monolithic bar of pure silicon. By summer 1958, Kilby had fabricated the first working semiconductor circuit, a simple phase shift oscillator.

The advancement in late 1958 by the U.S. Air Force and Westinghouse of a bold approach to the achievement of integrated electronics under the name of "molecular electronics" must also be recognized as one of the key initiating events. The development effort under this program began in early 1959.

THE ULTIMATE PERVASIVE CHARACTER OF ELECTRONICS

In assessing the impact of the technology of these and other history-making achievements, it is pertinent to take a long look at electronics. A particularly illuminating statement by Dr. Daniel E. Noble was published in Alfred Cook's editorial column in *Electronic News* on June 1, 1964. "For years," wrote Dr. Noble, "I have been referring to all electronics as generic art, and it should be apparent to every observer that the generic character of the art is extending rapidly to a point where the electronic applications are either found now or they will be found in the future in almost every pattern of activity associated with our scientific culture." Like Dr. Noble, I, too, have endeavored to express the same concept by describing electronics as pervasive in character and pointing out that as electronics fulfilled its inherent capabilities and pervaded all aspects of our society, it would be increasingly difficult to identify, in a really meaningful way, an electronics industry. The effectiveness with which we have been able to apply electronics and, in consequence, its pervasiveness have been limited by the inadequacies of our engineering skills and available tools. Obviously, a major objective of our technical effort has been to eliminate these limitations, and revolutions such as those cited have usually followed hard after each success. It now seems highly probable to me that integrated electronics will be the most successful of our accomplishments in removing limitations and, indeed, in a sense, may introduce a terminal phase during which electronics will pervade all segments of our society where it has pertinence.

In evaluating why this may be so and what it will take to make it so, it will be useful to view electronics from two different aspects:

1. The first is inward-looking and includes the science, the engineering and the art internal to electronics. Here are the knowledge and the tools used to create and make the materials, the devices and all of that myriad of circuitry we use to accomplish functions electronically.

2. The second is outward-looking and includes the applications of electronics, such as use of electronics to create machines to detect enemy aircraft or missiles, bring entertainment to the home, assist in the control of inventories, process payrolls, or control the refining of petroleum.

As engineers engaged in creating, making and marketing useful products and services for our society, we must recognize that we perform a useful function only as our efforts are displayed eventually in this second category, realizing simultaneously that these can be built only upon the inward-looking knowledge and the tools internal to electronics.

To say with Dr. Noble that electronics is a generic art, or with me that electronics is inherently pervasive, is simply to say that the basic knowledge and the tools of electronics are so pertinent to the needs of our kind of society that the products and services which are the result of the knowledge and tools have nearly unlimited usefulness and can contribute in a major way across our entire social structure.

BARRIERS TO OVERCOME IN ACHIEVING PERVASIVENESS

Yet, in spite of the pertinence of the knowledge and tools, there have been very fundamental limitations to our applying this knowledge and these tools as broadly as they justify and realizing the inherent power and full pervasiveness of electronics. Some of the most harassing have been:

1. The limitation of reliability
2. The limitation of cost
3. The limitation of complexity
4. The limitation imposed by the specialized character of and relative sophistication of the science, engineering and art of electronics.

The limitations are, of course, interrelated. Cost is obviously affected by the need for high reliability and necessarily complex solutions. Conversely, the more complex the solution required, the greater the likelihood that reliability and/or cost will become a controlling limitation. Such solid-state devices as transistors and diodes have certainly led the way to marked improvement in reliability, but they have hardly eliminated complexity. The solutions we have achieved still have a relatively high enough cost to inhibit the application of electronics in those broad areas which we customarily describe as the industrial and consumer sectors of our economy. So far as the fourth limitation is concerned, electronics is indeed a sophisticated branch of engineering and as such it has required highly skilled practitioners. Yet the very sophistication called for inevitably limits the rate at which electronics can pervade our society. For electronics to be truly pervasive, it must be readily and commonly used by the mechanical engineer, the chemical engineer, the civil engineer, the physicist, the medical doctor, the dentist, the banker, the retail merchant, and by the average citizen in broader ways than just for bringing entertainment to his home. Electronics cannot be truly pervasive unless such persons whose needs call for the powerful tools of electronics are capable of using them. It hardly seems feasible to suggest that all these highly skilled practitioners in other professions must also become skilled in the internal complexities of ours. The problem is considerably simplified, however, if the electronics skills which they require are limited to the comprehension and specification of the input and output parameters of the electronic functions they need. And, it is exactly here that integrated electronics may prove to remove a large percentage of these communication limitations.

THE SELF-HELP FEATURES OF THE NEW ELECTRONICS

The contributions integrated electronics is likely to make in removing limitations in the categories of reliability, cost and complexity are also impressive. Indeed, because integrated electronics seems to me to have a high probability of removing an appreciable percentage of the limitations in all four categories, I believe it may bring the total of these limitations to a critical level and, hence, initiate the terminal phase in which electronics contributes in truly vital ways to all segments of our society.

APPLICATIONS TO DATE

There are numerous applications which demonstrate that the limitations are indeed being removed. One is another historic first: the first complete equipment application of semiconductor integrated circuits was the data processor developed by Texas Instruments in 1961 under the sponsorship of the Manufacturing Technology Laboratory of the U.S. Air Force. Other real landmarks are three programs initiated in 1962: the Minuteman Guidance System (Autonetics Division, NAA), the Apollo Guidance Computer (MIT, Raytheon), and the Central Data Processor of the W2F Airplane (Data Systems Division, Litton Industries). All of these programs are demonstrating important gains over their conventional discrete component counterparts.

Early digital integrated circuits lagged behind their transistor circuit equivalents by a year or two in comparable state of art performance, primarily because of capacitance between the diffused regions and the substrate and the longer and higher resistivity paths in the collectors of all top contact transistor structures. The use of epitaxial layers within the past year has minimized this design problem. Today, developments in dielectric isolation are minimizing parasitic capacitance. As a consequence, performance equivalent to that obtainable with discrete transistor circuitry, particularly in terms of propagation time, may now be expected. It would also appear probable that the combination of a large number of digital circuits on a single semiconductor substrate promises real reduction in transmission time between circuits so that the net effect will be faster over-all performance for these complex integrated circuits than is likely for discrete transistor circuitry.

APPLICATIONS TO NON-DIGITAL (LINEAR OR ANALOG) SOURCES OF INFORMATION

To achieve the broad impact on electronics which I forecast, integrated circuitry must, of course, become as applicable to the creation of electronic functions we customarily think of as linear as it already is for those we describe as digital. It is true that the development of linear circuits has moved more slowly than digital due primarily to the limited range of passive components that can so far be formed in the semiconductor material. Nevertheless, differential amplifiers have been applied to the Minuteman II Guidance System (Autonetics Division, NAA) and the ASN-44 Inertial Platform (Guidance and Control Division, Litton). Of significance, too, is the introduction by Zenith early in 1964 of a hearing aid using a TI integrated circuit audio amplifier. I am convinced, however, that the utilization of epitaxial and dielectric layers plus the exploitation of new phenomena will allow the broad use of integrated circuitry for linear functions. As a matter of fact, achieving specific electronic functions through the exploitation of new phenomena rather than by simply recreating within the solid substrate approximately the same circuit as that utilized with discrete components suggests that we will be able to reach beyond the effectiveness of discrete componentry.

RECENT GROWTH

To judge how integrated electronics will affect us, and both what needs to be done and the likely course of events, requires that we endeavor to measure the growth of integrated circuitry over recent years and forecast the probable growth for a reasonable span of future years.

Actually, as was almost inevitable because of the inheritance of semiconductor technology and applications, the production volume of integrated circuits has grown more rapidly than did that for transistors during a comparable period. Note that the contrast in growth rates is even more pronounced when the data in the chart are adjusted to reflect the average number of transistors displaced per integrated circuit so that comparable penetration of equipment applications is displayed.

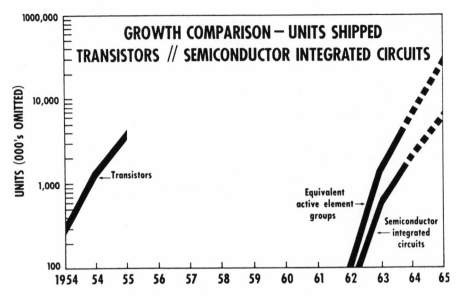

FIG. A1. Haggerty's diagram comparing the rate of growth of discrete transistors shipped by Texas Instruments in the mid-1950s to the corresponding rate of growth of equivalent active element groups and integrated circuits in the early production years of the 1960s. The rates of growth for Texas Instruments in this period exceeded the average rates of overall growth predicted by Gordon Moore a year later, but Texas Instruments was in a preferred position at this early period. (Courtesy of Texas Instruments Incorporated.)

ATTEMPTED FORECAST

Any forecast must start with the recognition that there are no well-defined statistics available on the size and composition of the total electronics circuit market. Elsewhere (see *Spectrum,* June 1964), I have reported on one study assessing this electronics circuit market and the pressures it suggests for conversion from discrete componentry to integrated circuitry. This study based its statistical approximations on an estimate of the number of active element groups in various kinds of electronic circuitry. An active element group was arbitrarily defined as an active element plus its associated circuitry including diodes, resistors, capacitors, relays, inductors, coils, connectors, printed circuit boards, etc. The number of active elements was determined on the basis of total transistors, controlled rectifiers, 37.5 percent of all other rectifiers (reflecting their use in bridges) and 170 percent of the number of receiving tubes (reflecting equivalent usage of dual element

1973 POTENTIAL	GOVT.	INDUSTRIAL	CONSUMER	TOTAL
ELECTRONIC EQUIPMENT SALES	$10.7B	5.7B	3.7B	20.1B
COST OF CIRCUITS USED IN EQUIPMENT	3.1B	1.8B	.9B	5.8B
VALUE OF CONVENTIONAL CIRCUITS POTENTIALLY REPLACEABLE BY INTEGRATED CIRCUITS (AT BREAKEVEN)	1.8B	.7B	.4B	2.9B
SAVINGS IF INTEGRATED CIRCUITS COST 50% LESS THAN CONVENTIONAL CIRCUITS				
AT CIRCUIT LEVEL	.9B	.3B	.2B	1.4B
AT CHASSIS FABRICATION LEVEL	.5B	.1B	—	.6B
TOTAL POTENTIAL SAVINGS	$1.4B	.4B	.2B	2.0B

FIG. A2. Haggerty's too conservative estimate in 1964 of what overall sales might be a decade later. The values are in billions (10^9) dollars. A year later, in 1965, the situation was somewhat clearer and Gordon Moore was prepared to be bolder, even before the invention of the microprocessor. (Courtesy of Texas Instruments Incorporated.)

tubes) going into new equipment. These data suggest that the total number of active element groups required for all electronic circuitry in 1963 was 715 million units with average value of $5.04 each. Thus, 1963 total circuit costs are estimated as approximately 25 percent of equipment sales for a total of $3.6 billion.

In any effort to project the economic impact of integrated circuits for the next decade, the variables involved are so many and the rate of change in growth of technology so high that fundamental parameters in the assessment are almost certain to be not just wrong but strikingly so. Nevertheless, the study used the same techniques to generate a model of the pressures for conversion to integrated circuitry in 1973.

BASIC REQUIREMENTS FOR FUTURE SUCCESS

The basic requirements to ensure that electronics enters this terminal phase of pervasiveness, I believe, are threefold:

1. A relatively concentrated, highly automated industrial complex which supplies integrated circuitry and closely related compatible discrete componentry to the rest of the electronics industry and to industry in general must exist. Only a few organizations (perhaps five) will supply 90 percent or more of total industry needs, for this will be a heavily capitalized industry with elaborate computer-controlled processing plants necessary to provide the great flexibility essential to produce the wide variety of integrated circuits needed to fulfill 50 percent or more of all electronic function requirements. In essence, this will be a basic materials segment of the electronics industry with the integrated circuits it produces as the basic materials used by the much larger total electronics industry to satisfy the needs of its customers. In a very real sense (although one must not pursue the analog too far), the integrated circuit producers will be to the rest of the industry as the producers of steel are to the automotive industry, the producers of copper are to the electrical industry, or the producers of aluminum to the myriad of organizations which use that material as a basis for their products.

2. This integrated circuits industry must have established a common language for the input and output parameters which specify its products. It will have created a wide variety of computer programs, which will have replaced conventional engineering handbooks as we know them today and truly allow the user of these basic electronic materials, integrated circuits and compatible discrete components, to design the required electronic functions by computer in terms of the input and output parameters available and specified.

3. A very large number of organizations, probably many more than today, will utilize these basic electronic materials to solve their own and their customers' problems. These organizations will exist in all sectors of our society and will be able to utilize the highly specialized and highly concentrated integrated circuits industry as a substitute for the kind of sophisticated electronics skill described above as the fourth limitation. This will have been made possible by the myriad of computer programs which will allow design by computer through the specification via common language of input and output parameters. A much larger proportion than today of our highly talented electronics engineers will be able to devote their time to the application of electronics to meet the needs of our society rather than to looking inward at electronics itself.

All three of these conditions are interdependent, but the first undoubtedly is the primary requisite to achieve the other two. The reasoning behind

the belief that there must be a highly concentrated integrated circuits industry is this.

In future years, if the requisite technical and economic levels are to be
reached, a very large proportion of our technical effort must go into improving materials and processing technology, trending toward continuous
process flow with real time feedback and control. As compared to the batch
processes inherent in present semiconductor technology, continuous process flow offers the potential for very large gains in cost reduction, quick
reaction time and economic short-run capability. The resultant facilities
must be highly automated and provide adequate coupling into the input
and output parameter computer programs called for above. The trend toward larger arrays, particularly for logic functions, has already been noted.
If a $750 million market for integrated circuitry does develop by the early to
middle 1970s, as I have suggested, perhaps three-quarters of the 780 million
active electronic groups (AEGs) estimated as replaceable by integrated circuitry will probably have been so replaced. At an average of 10 AEGs per
integrated circuit, the total volume would be 58.5 million integrated circuits.
While this is a large-volume market, in terms of quantity it falls far short of
current annual production of 300 million transistors and slightly *less than
300 million receiving tubes.* Thus, it seems very doubtful that the difficult
materials and processing research and development, and the highly complex
capital equipment required to make these real-time computer-controlled
automated factories feasible, can be economic for a total annual volume of
58.5 million units if more than five major producers are involved. For a plant
of sufficient scale and complexity to handle 10 percent or more of the total
volume available on an economic basis, it is not at all unlikely that capital
equipment with costs measured in terms of tens of millions of dollars would
be required.

In any event, integrated electronics promises to increase both the rate of
change within our own industry was well as the pervasiveness of electronics
as a whole. Surely the challenges inherent in bringing these three requisite
conditions about match or exceed anything we have faced previously in our
profession and offer promise that all of us in electronics may in the future be
able to contribute even more effectively to the improvement of our society
and through that improved contribution gain the personal rewards and sat-
isfactions which accompany it.

Computers will be more powerful, and will be organized in completely
different ways. For example, memories built of integrated electronics may

be distributed throughout the machine instead of being concentrated in a central unit. In addition, the improved reliability made possible by integrated circuits will allow the construction of larger processing units. Machines similar to those in existence today will be built at lower costs and with faster turn-around.

GORDON MOORE'S FORECAST
(1965)

The following paragraphs are reproduced from the article "Cramming More Components onto Integrated Circuits," by G. E. Moore, *Electronics,* April 19, 1965, p. 114. Several paragraphs in the original version have been omitted and subheadings have been modified.

PRESENT AND FUTURE

By integrated electronics, I mean all the various technologies which are referred to as microelectronics today as well as any additional ones that result in electronics functions supplied to the use as irreducible units. These technologies were first investigated in the late 1950s. The object was to miniaturize electronics equipment to include increasingly complex electronic functions in limited space with minimum weight. Several approaches evolved, including microassembly techniques for individual components, thin-film structures and semiconductor integrated circuits.

Each approach evolved rapidly and converged so that each borrowed techniques from another. Many researchers believe the way of the future to be a combination of the various approaches.

The advocates of semiconductor integrated circuitry are already using the improved characteristics of thin-film resistors by applying such films directly to an active semiconductor substrate. Those advocating a technology based upon films are developing sophisticated techniques for the attachment of active semiconductor devices to the passive film arrays.

Both approaches have worked well and are being used in equipment today.

NEW ELECTRONICS FIRMLY ESTABLISHED

Integrated electronics is established today. Its techniques are almost mandatory for new military systems, since the reliability, size and weight required by some of them is achievable only with integration. Such programs as Apollo, for manned moon flight, have demonstrated the reliability of integrated electronics by showing that complete circuit functions are as free from failure as the best individual transistors.

Most companies in the commercial computer field have machines in design or in early production employing integrated electronics. These machines cost less and perform better than those which use "conventional" electronics.

Instruments of various sorts, especially the rapidly increasing numbers employing digital techniques, are starting to use integration because it cuts costs of both manufacture and design.

The use of linear (analog) integrated circuitry is still restricted primarily to the military. Such integrated functions are expensive and not available in the variety required to satisfy a major fraction of linear electronics. But the first applications are beginning to appear in commercial electronics, particularly in equipment which needs low-frequency amplifiers of small size.

RELIABILITY COUNTS

In almost every case, integrated electronics has demonstrated high reliability. Even at the present level of production—low compared to that of discrete components—it offers reduced systems cost, and in many systems improved performance has been realized.

Silicon is likely to remain the basic material although others will be of use in specific applications. For example, gallium arsenide will be important in integrated microwave functions. But silicon will predominate at lower frequencies because of the technology which has already evolved around it and its oxide, and because it is an abundant and relatively inexpensive starting material.

COSTS AND CURVES: MOORE'S LAW

Reduced cost is one of the big attractions of integrated electronics, and the cost advantage continues to increase as the technology evolves toward the production of larger and larger circuit functions on a single semiconductor substrate. For simple circuits, the cost per component is nearly inversely proportional to the number of components, the result of the equivalent piece of semiconductor in the equivalent package containing more components. But as components are added, decreased yields more than compensate for the increased complexity, tending to raise the cost per com-

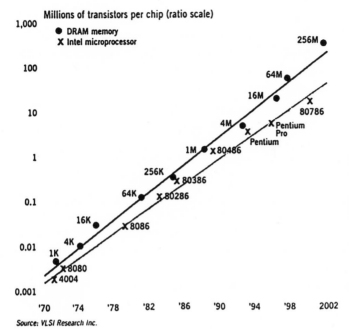

FIG. B1. One valida-
tion of Moore's law.
This diagram shows
the rate of growth of
the production of
DRAMs and of Intel
microprocessors
over three decades.
(Courtesy of Forbes
Incorporated and
VSLI Research
Incorporated.)

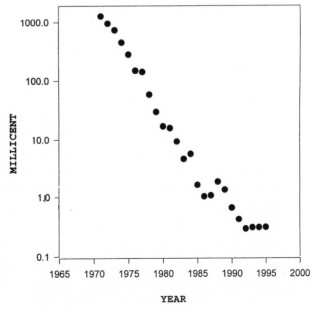

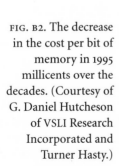

FIG. B2. The decrease
in the cost per bit of
memory in 1995
millicents over the
decades. (Courtesy of
G. Daniel Hutcheson
of VSLI Research
Incorporated and
Turner Hasty.)

ponent. Thus there is a minimum cost at any given time in the evolution of the technology. At present, it is reached when 50 components are used per circuit. But the minimum is rising rapidly while the entire cost curve is falling. If we look ahead five years, a plot of costs suggests that the minimum cost per component might be expected in circuits with about 1,000 components per circuit (providing such circuit functions can be produced in moderate quantities). In 1970, the manufacturing cost per component can be expected to be only a tenth of the present cost.

The complexity for minimum component costs has increased at a rate of roughly a factor of two per year. Certainly over the short term this rate can be expected to continue, if not increase. Over the longer term, the rate of increase is a bit more uncertain, although there is no reason to believe it will not remain nearly constant for at least 10 years. That means by 1975, the number of components per integrated circuit for minimum cost will be 65,000.

I believe that such a large circuit can be built on a single wafer.

TWO-MIL SQUARES

With the dimensional tolerances already being employed in integrated circuits, isolated high-performance transistors can be built on centers two thousandths of an inch (fifty microns) apart. Such a two-mil square can also contain several kilohms of resistance or a few diodes. This allows at least 500 components per linear inch or a quarter million per square inch. Thus, 65,000 components need occupy only about one-fourth a square inch.

On the silicon wafer currently used, usually an inch (2.54 centimeters) or more in diameter, there is ample room for such a structure if the components can be closely packed with no space wasted for interconnection patterns. This is realistic, since efforts to achieve a level of complexity above the presently available integrated circuits are already underway using multilayer metalization patterns separated by dielectric films. Such a density of components can be achieved by present optical techniques and does not require the more exotic techniques, such as electron beam operations, which are being studied to make even smaller structures.

INCREASING THE YIELD: WILL PACKAGING COSTS DOMINATE?

There is no fundamental obstacle to achieving device yields of 100%. At present, packaging costs so far exceed the cost of the semiconductor structure itself that there is no incentive to improve yields, but they can be raised as high as is economically justified. No barrier exists comparable to the thermodynamic equilibrium considerations that often limit yields in chemical reactions; it is not even necessary to do any fundamental research or to replace present processes. Only the engineering effort is needed.

In the early days of integrated circuitry, when yields were extremely low, there was such incentive. Today, ordinary integrated circuits are made with yields comparable with those obtained for individual semiconductor devices. The same pattern will make larger arrays economical, if other considerations make such arrays desirable.

HEAT PROBLEM

Will it be possible to remove the heat generated by tens of thousands of components in a single silicon chip?

If we could shrink the volume of a standard high-speed digital computer to that required for the components themselves, we would expect it to glow

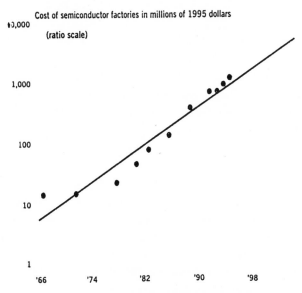

FIG. B3. Another, complicating, aspect of Moore's law, sometimes referred to as Moore's second law: The geometrical rise in the cost of fabricating units. Daniel Hutcheson, of VSLI Research, which carries on continuous studies of trends in semiconductor technology, had begun calling attention to the rapidly rising costs of fabricating units in the 1980s. (Courtesy of Forbes Incorporated and VSLI Research Incorporated.)

brightly with present power dissipation. But it won't happen with integrated circuits. Since integrated electronic structures are two-dimensional, they have a surface available for cooling close to each center of heat generation. In addition, power is needed primarily to drive the various lines and capacitances associated with the system. As long as function is confined to a small area on a wafer, the amount of capacitance which must be driven is distinctly limited. In fact, shrinking dimensions on an integrated structure makes it possible to operate the structure at higher speed for the same power per unit area.

DAY OF RECKONING

Clearly, we will be able to build such component-crammed equipment. Next, we ask under what circumstances we should do it. The total cost of making a particular system function must be minimized. To do so, we could amortize the engineering over several identical items, or evolve flexible techniques for the engineering of large functions so that no disproportionate expense need be borne by a particular array. Perhaps newly devised design automation procedures could translate from logic diagram to technological realization without any special engineering.

It may prove to be more economical to build large systems out of smaller functions, which are separately packaged and interconnected. The availability of large functions, combined with functional design and construction, should allow the manufacturer of large systems to design and construct a considerable variety of equipment both rapidly and economically.

LINEAR (ANALOG) CIRCUITRY

Integration will not change linear systems as radically as digital systems. Still, a considerable degree of integration will be achieved with linear circuits. The lack of large-value capacitors and inductors is the greatest fundamental limitation to integrated electronics in the linear area.

By their very nature, such elements require the storage of energy in a volume. For high Q (narrow resonance) it is necessary that the volume be large. The incompatibility of large volume and integrated electronics is obvious from the terms themselves. Certain resonance phenomena, such as those in piezoelectric crystals, can be expected to have some applications for tuning functions, but inductors and capacitors will be with us for some time.

The integrated radio-frequency amplifier of the future might well consist of integrated stages of gain, giving high performance at minimum cost, interspersed with relatively large tuning elements.

Other linear functions will be changed considerably. The matching and tracking of similar components in the integrated structures will allow the design of differential amplifiers of greatly improved performance. The use of thermal feedback effects to stabilize integrated structures to a small fraction of a degree will allow the construction of oscillators with crystal stability.

Even in the microwave area, structures included in the definition of integrated electronics will become increasingly important. The ability to make and assemble components small compared with the wavelengths involved will allow the use of lumped parameter design, at least at the lower frequencies. It is difficult to predict at the present time just how extensive the invasion of the microwave area by integrated electronics will be. The successful realization of such items as phased-array antennas, for example, using a multiplicity of integrated microwave power sources, could completely revolutionize radar.

NAME INDEX

SUBJECT INDEX

Acceptor levels, 62

Activation energy, 50, 51, 127

Alferov letter, Soviet radar history, 109

Alkali halides: and F centers, 53; and ionic
conductivity, 54

Alsos Mission, 95

Amateur radio, 41

American Telephone and Telegraph
(AT&T): and acquisition of triode
vacuum tube, 153; and the audion, 40;
breakup of, and Judge H. H. Greene,
151; and Western Electric Company, 151

Ampex Corporation, magnetic memories,
215

Amplifier, 39

Amplitude modulation, 42

Analog (linear) processes, 197

Analog to digital processing, seismic data,
222, 227n

Anomalous Hall effect, holes, 60

Antenna, dipole, 9, 17

Antenna arrays, 40, 41

Armstrong–De Forest lawsuits, 40

Army Signal Corps, radar, 125

Arrhenius (Boltzmann) plot, 50

AT&T. *See* American Telephone and
Telegraph

Atmospheric ionization layers, 90

Atomic Research Establishment
(Harwell), and Skinner, 118

Audion, 39, 40

Bakelite, 30

Band gap: diamond, 59; silicon carbide,

38n; silicon and germanium, 127; Seiler,
Gudden, 98

Band structure: diamond, Kimball, 57, 58,
71; sodium metal, Slater, 57, 70; types,
58, 70, 71

Band theory of solids, 56; Strutt, Bloch,
Wigner, and Seitz, 56

Bardeen and Shockley, new careers,
186–89

Bardeen patent, field-effect transistor, 173

Barium oxide coating, 27

Bell Telephone Laboratories, 151; discovery of p-n junction, 156; radio echoes,
125; restriction on visitors, 57; retro-
spective, 151–56; visits to, 126

Bidwell's measurements, germanium, 127,
128

Binary-digital processes, 197

Bipolar junction transistor, 156, 171–74,
180; diagram, 180; Shockley, 171; variations, 172

Bipolar point-contact transistor, 168–69,
180; diagram, 180

Blocking layer, 51, 78, 82

Blocking mode, metal-semiconductor
junction, 169

Boltzmann (Arrhenius) plot, 50

Boone's account of microcontroller, 228,
236n

Boron, addition to silicon, 129

British General Electric Company (GEC),
119

British radar, 113–23

British Thomson-Houston (BTH), 116

Torrey committee, centralized planning, 131

Torrey-Whitmer book, 143–45

Transatlantic reception, Marconi, 22

Transcontinental telephone, 153

Transistor: mesa, 202, 207; naming of, 169; Nobel prize, 187, 191; planar, 202, 207

Traveling wave tube, 124

TRE. *See* Telecommunications Research Establishment

Trichlorosilane, pure silicon, 177

Triode vacuum tube: acquisition by AT&T, 153; Langmuir, 153; repeater stations, 153

Triode, semiconductor, French, 103n, 174

Tube de Limaille, 21

Tungsten-silicon diodes, state of art, World War II, 144

University of Bristol, Skinner, 116

University of Illinois, Ordvac and Illiac, 198

University of Pennsylvania, 126, 128; ENIAC, 197

University of Tokyo, 149

U.S. Army Signal Corps: technical pamphlet (1918), 29, 36; wireless, 29, 36

U.S. Patent Office, award to Boone (1996), 229

User-friendly systems, Apple Computer, Steven Jobs, 242

Vacuum technology, 47n; Dushman, Gaede, 47n

Vacuum tube: amplifier, 39; audion, 39; Morton, 130; oscillator, 39; period of dominance, 29; triode, De Forest, 39; limitation for computers, 210n

Vacuum tube alternative, 130

Vacuum tube diode, 27

Vacuum tube era, 39–43

Vacuum tube mixer, 126

Visionary forecasts, of Haggerty and Moore, 220–24, 251, 261

Wave mechanics, 55

Western Electric Company, AT&T, 151

Westinghouse Research Laboratory: Condon, 153; Angello, 132

West Street Building, photograph, 158

Wireless telegraphy, 8, 20; and Marconi, 20–21, 22, 23, 24, 25

Work function, 77

World marketplace, role of electronic communications in, 224

Wright Air Development Command, financing of integrated circuit, 213

Yield, improvement, diagram, 247

Zone purification: Pfann, 198, 204; equipment, diagram, photograph, 204

FREDERICK SEITZ is president emeritus of Rockefeller University, past president of the National Academy of Sciences, and an honorary member of the National Academy of Engineering. His involvement in research in solid-state physics began in 1932 as a graduate student at Princeton University, working with Edward U. Condon and Eugene P. Wigner. He has served as president of the American Physical Society.

NORMAN G. EINSPRUCH is senior fellow in science and technology and professor and chair of the Department of Industrial Engineering at the University of Miami, where he served as dean of the College of Engineering for thirteen years. His first career was in the electronics industry, where he served as director of the Central Research Laboratories and as assistant vice president of Texas Instruments Incorporated.